DATE DUE

AG 5 00		
OC 8 01		

DEMCO 38-296

TESTING AND BALANCING
HVAC
AIR AND WATER SYSTEMS

Second Edition

TESTING AND BALANCING HVAC AIR AND WATER SYSTEMS

Samuel C. Monger

Second Edition

Published by
THE FAIRMONT PRESS, INC.
700 Indian Trail
Lilburn, GA 30247

TH 7015 .M66 1995

Monger, Sam, 1946-

Testing and balancing HVAC
 air and water systems

:ion Data

ater systems / Sam Monger.

Includes index.
ISBN 0-88173-210-9
    1. Heating--Equipment and supplies--Testing. 2. Ventilation--Equipment
and supplies--Testing. 3. Air conditioning--Equipment and
supplies--Testing. 4. Balancing of machinery. I. Title.
TH7015.M66   1995   697--dc20   95-17282
                                                     CIP

*Testing And Balancing HVAC Air And Water Systems By Sam Monger,
Second Edition.*

Published by The Fairmont Press, Inc.
700 Indian Trail
Lilburn, GA 30247

Printed in the United States of America

10 9 8 7 6 5 4 3 2 1

ISBN 0-88173-210-9   FP

ISBN 0-13-462276-6   PH

While every effort is made to provide dependable information, the publisher, authors, and
editors cannot be held responsible for any errors or omissions.

Distributed by Prentice Hall PTR
Prentice-Hall, Inc.
A Simon & Schuster Company
Upper Saddle River, NJ 07458

Prentice-Hall International (UK) Limited, London
Prentice-Hall of Australia Pty. Limited, Sydney
Prentice-Hall Canada Inc., Toronto
Prentice-Hall Hispanoamericana, S.A., Mexico
Prentice-Hall of India Private Limited, New Delhi
Prentice-Hall of Japan, Inc., Tokyo
Simon & Schuster Asia Pte. Ltd., Singapore
Editora Prentice-Hall do Brasil, Ltda., Rio de Janeiro

CONTENTS

FOREWORD

This book is for anyone interested in testing and balancing HVAC systems. For the novice and the experienced test and balance technician, it's a field reference book of procedures, equations and useful conversion tables. For anyone interested in getting into testing and balancing or who's presently a balancing technician, it's a text for learning more about balancing and HVAC systems. For the engineer or building manager, it's a text for teaching test and balance, writing more effective specifications, and understanding the principles of balancing for improved communications with balancing technicians and ensuring that the system is being properly balanced.

This book is used in test and balance self-study courses, in-house training programs and as a supplemental text in college and technical school courses on fluid dynamics and air conditioning. It is arranged in four parts. Part 1 has a general balancing procedure, and specific air balancing procedures for both constant air volume and variable air volume systems. New in this section is a chapter on testing and balancing fume hoods. Part 2 is water balancing procedures using flow meters, system components and temperatures. Part 3 goes into more depth on types of HVAC systems, system components, balancing devices, and balancing instrument usage and care. Part 4 is equations and conversion tables. Use this section to write in your own equations, CEFAPPs, correction factors, etc.

As always, comments, criticisms and war stories are welcome. I commend you for your pursuit of knowledge and excellence in the HVAC&R industry, and your commitment to provide the highest quality of service. Good Luck!

Samuel C. Monger
San Diego, CA

INTRODUCTION

Balancing can be interesting and challenging. But, it can also be difficult and frustrating because air systems will sometimes lack balancing dampers and frequently, water systems will have neither flow meters nor balancing valves. What do you do in these cases? Well, first you tell someone about the problem. Try to get balancing devices and in water systems, flow meters installed. Physically, putting in the dampers in the ductwork will be easy, installing valves and flow meters in piping will be more difficult. However, the most difficult task will probably be getting someone to pay for them. The next problem is getting enough straight duct or pipe to take accurate flow quantity measurements. What do you do? You take as many varied readings as necessary to satisfy yourself that you can honestly say what the flow quantities are and how the system is operating. In other words, you do the best that you can with what you have, realizing that sometimes systems can't be balanced to design quantities, temperatures, humidities, flow patterns, etc.

Beside being aware of the problems associated with proportioning flow quantities, you should also be aware of conditions which (1) hamper the flow of air and water, such as fan and pump system effect, restrictive ductwork (fittings, extensive lengths of flex duct, etc.) and restrictive pipe fittings, (2) increase the loss of thermal energy such as uninsulated ducts or pipes, large aspect ratios, and leaking dampers or valves, (3) reduce proper air mixing in the unit or in the space, (4) lessen control of the space because of improper location or calibration of sensors, and (5) increase the use of energy, such as fans or pumps dampered or valved down instead of reducing fan speed or pump impeller size. Inform owners, engineers or contractors of the problems with equipment, duct and piping design or installation.

There are two basic methods of balancing. One method is balancing from the fan or pump out and the other is balancing from the terminals back to the fan or pump. Balancing can be further broken down into (1) experience balancing or (2) proportional balancing. Experience balancing means that you must work in the field for many years learning different types of systems and from experience gain an insight for which damper or valve to cut and where the air or water will go. I've chosen to explain in this book proportional balancing from the terminals back to the prime mover.

This book is written from my field experience in balancing and evaluating HVAC systems and teaching test and balance. It was during a job at a nuclear power plant some years ago that I first started using the proportional method that I've outlined in this text. Because we couldn't get enough experienced balancing technicians for the project, I began instructing sheet metal mechanics with little or no balancing background at the plant. I found that (1) the proportional method was learned quicker (obviously the experience method takes years), and (2) the balancing was done faster. I also found that proportional balancing is a better method than experience balancing because (1) it follows a scientific approach with predictable results, (2) it establishes a starting point and allows others with proportional balancing experience to come in and take over the balancing when the need arises, (3) large systems are balanced quicker, and (4) the least resistance is imposed on the system while delivering air or water to the terminals at design or at the highest possible flow quantities. I've continued to teach this method in my classes, seminars and in-house training programs with great success. I'm sure it'll work for you as well.

In the chapters on proportional balancing, I've used the term percent of design (%D). This is meant for new construction where balancing is done by an agency according to the fluid quantities specified on the contract documents. These numbers are based on expected heating and cooling loads. However, %D can also stand for percent of desired flow quantity. This is used for in-house balancing or balancing to owners' specifications and is based on the actual space operating loads, location of occupants and equipment and comfort conditions. This is known as a comfort balance.

What are some benefits of proper testing and balancing? Well, for occupants, there's increased comfort resulting from proper air movement and the correct number of air changes; for owners, lower operating costs. The life of the HVAC equipment is extended and there can be large savings in energy usage when the flow quantities of systems over on air or water are lowered by reducing the speed of the fan or the size of the pump impeller. Are there benefits for the facility manager or maintenance engineer? Yes, many. Testing and balancing is a commissioning service uncovering problems in operation, installation and design. Also, when the balancing technicians are finished with the system they'll be able to supply an updated set of "as-built drawings" and a table of equipment. The design engineer and the balancing technician can also benefit. Through review of the balance reports and by taking a synergistic approach to consult with each other to clarify and solve problems, the design engineer and balancing technician can work together to improve system design and balancing techniques and practices. This brings us full circle. Who benefits from proper testing and balancing? Everyone.

PART 1

Chapter 1
General Balancing Procedure

Balancing HVAC air and water systems calls for a systematic approach. While there is no one specific balancing procedure, the following is a general procedure which can be applied to all systems.

1. Do the preliminary office work.
   a. Gather plans and specifications.
   b. Prepare report forms.
2. Do the preliminary field inspection.
   a. Inspect the job site.
   b. Inspect the distribution system.
   c. Inspect the equipment.
3. Make initial tests on all fans and pumps applicable to the system being balanced.
4. Balance and adjust the distribution system.
5. Adjust the fan or pump as needed.
6. Take final readings.
7. Complete reports.

PRELIMINARY OFFICE WORK

First, gather all applicable plans and specifications to include contract drawings, shop drawings, "as-built" drawings, schematics, manufacturers' catalogs showing equipment description and capacities, manufacturers' data and recommendations on testing their equipment, equipment performance curves, etc. Not all items will be available, but the more information that you can get the better your understanding will be of the system and its components.

Next, study the plans and specifications to become familiar with the system and the design intent. To help you with becoming familiar with the system, color in the drawings. This will also make them easier to read. The supply air or water might be colored blue, the return air or water - red, the exhaust systems - green, etc. Then, label all pieces of equipment. SF for supply fan, EF for exhaust fan, Ch-1 for chiller number 1, CHWP-2 for chilled water pump number 2, etc. Next, starting from either the end or the beginning of the system, number all the terminal devices such as coils, boxes, diffusers, etc. On the air side, circle all volume dampers, splitters, automatic dampers, fire dampers, etc. On the water side, circle all balancing valves, flow meters, automatic control valves, etc.

After becoming familiar with the system, a determination of the method of balancing will need to be made. This will include a determination of what readings will be taken and where, and what instruments will be needed.

Finally, make out the report forms for each system being tested. Report forms will include equipment test sheets and balancing sheets for the distribution system. Depending on the size and complexity of the project, general information sheets and summary sheets may also be needed. I also recommend that a schematic drawing of each system be made. Schematics may be very detailed showing the central system with pressure and temperature drops across components, duct and pipe sizes, terminal devices, balancing valves and dampers, required and actual flow quantities, etc. But, at the very least, the schematic should show the numbered terminal devices with desired flow quantities and the location of volume dampers and valves.

Do as much paperwork in the office as possible where lighting, work tables, access to reference materials etc. are more favorable.

PRELIMINARY FIELD INSPECTION

After the reports are prepared, inspect the job site to see that the building is architecturally ready to be balanced. For instance, are all the walls, windows, doors, and ceilings installed? If the conditioned space isn't architecturally sealed, abnormal pressures and temperatures will adversely affect the system balance.

Next, walk the air and water distribution systems to ensure that they are intact, and aren't missing components such as dampers, valves, pressure and temperature taps, coils, terminal boxes, diffusers, grilles, etc.

Lastly, inspect the equipment. Check that motors, fans, pumps, chillers, compressors, boilers, drives, etc. are mechanically and electrically ready. The following is a general list of items to be checked.

Air Side
1. Ductwork intact and properly sealed.
2. Ductwork leak tested.
3. Access doors installed and properly secured.
4. Ductwork installed according to the drawings and specification.
5. Ductwork free from debris.
6. Dampers, including fire and smoke dampers, installed and accessible.
7. Terminal boxes, reheat coils, electrical reheat, etc., installed, functional and accessible.
8. All other air distribution devices such as diffusers, etc., installed and functional.
9. Return air has an unobstructed path from each conditioned space back to the unit.
10. Filters clean and correctly installed.
11. Filter frame properly installed and sealed.
12. Coils cleaned and properly installed.
13. Drive components installed.
14. Sheaves properly aligned and tight on their shafts.
15. Belts adjusted for the correct tension.
16. Belt guard properly installed.
17. Automatic control dampers installed and functional.
18. Fan vortex dampers functional.
19. Fan housings installed and properly sealed according to the drawings and specifications.
20. Flexible connections installed properly.
21. Fan wheel aligned properly and adequate clearance in the housing.
22. Fan bearings lubricated.

Water Side
1. Strainers and piping free from debris, cleaned and flushed.
2. Construction strainer baskets replaced with permanent baskets.
3. System filled to the proper level and the pressure-reducing valve set.

4. Automatic and manual air vents properly installed and functional.
5. All air purged from the system.
6. Water in the expansion tanks at the proper level.
7. All valves, flow meters, and temperature/pressure taps installed correctly, accessible and functional.
8. Terminal coils installed and piped correctly and accessible.
9. Pumps properly aligned, grouted and anchored.
10. Vibration isolators properly installed and adjusted.
11. Flexible connections installed properly.

Boiler
1. All operating and safety settings for temperature and pressure are correct.
2. Pressure relief valve functional.
3. Boiler started and operating properly.

Chiller and Condenser
1. All operating and safety settings for temperature and pressure are correct.
2. Chiller and condenser started and operating correctly.

Electrical
1. Motors wired and energized.
2. Proper starter and overload protection installed.
3. Correct fuses installed.
4. Motors properly secured on their frames.
5. Motor bearings lubricated.

Controls
1. Controls complete and functional.

Chapter 2
Preliminary
Air Balancing Procedures

This chapter will explain preliminary balancing procedures including job site inspections, gathering motor nameplate information, thermal overload protection data, and drive information, and the initial tests required for balancing the air side of the HVAC system.

PRELIMINARY OFFICE WORK

Plans and Specifications

Gather all applicable plans and specifications including contract drawings, shop drawings, "as-built" drawings, schematics, automatic temperature control drawings, manufacturers' catalogs showing fan and terminal box description and capacities, and manufacturers' data and recommendations on testing fans, boxes, inlets and outlets. Also, try to get performance curves for fans, equipment operation and maintenance instructions and any previous air balance reports. Not all these items will be available for every system, but the more information that you can get, the better your understanding will be of the system and its components.

Study the plans and specifications to become familiar with the system and the design intent. Look for equipment, components or conditions that will change the air volume, shut down the system, affect the balance procedure or change the sequence of balancing the system. These items will be investigated during the preliminary field inspection. Items to look for:

1. Does the system have an adequate number of manual volume dampers for proper balancing? Each outlet should have a manual volume damper in the duct runout or takeoff leading to the outlet. Depending on the system, each branch duct and zone duct may also

need a manual volume damper.

2. Does the system have any features which may contribute to an unbalanced condition such as kitchen or lab fume exhaust fans that don't have adequate make-up air? Is the return system ducted or a non-ducted ceiling return plenum?

3. Will accessibility be a problem? Time delays may occur because of high ceilings, spline ceilings, equipment that is difficult to get to, or availability of limited access areas such as security areas, clean rooms, hotel rooms, meeting rooms, etc.

4. Will temporary heat requirements such as the need to put the system in a full heating or full cooling mode change the sequence of balancing?

For clarity and familiarization, color and label the drawings. Examples: supply air - blue, return air - red, exhaust systems - green, SF for supply fan, RF for return fan, AHU for air handling unit, etc. Then, number all the terminal devices (boxes, diffusers, etc.).

After familiarization with the system, make a tentative decision of the method of balancing. This will include a determination of what readings will be taken and where, if a correction for air density and air measurements will be needed because of altitude and/or temperatures, and what instruments will be needed. These items will be confirmed during the field inspection.

Report Forms

Prepare report forms for each system being tested. Report forms will include equipment test sheets and balancing sheets for the distribution system. Using the prints and specs complete all design data such as air quantities, fan information, motor information, and air distribution information. The test and balance report is a complete record of design, preliminary and final test data. It reflects the actual tested and observed conditions of all systems and components during the balance and lists any discrepancies between the specifications and the test data and possible reasons for the differences.

Included with the necessary test data sheets should be a separate sheet which contains a list of the various instruments used on the job. Before leaving for the job, verify that the instruments meet the calibration criteria stated in the balancing specifications or that the instruments have been calibrated within the last 6 months or to manufacturer's recommendations. Calibration can be done either by a certified calibration agency or by the test and balance agency if it has a sheltered set of instruments.

The total quantity of design supply air can be determined by adding up the air quantities for each supply outlet. Compare this total with the specified design capacity for the supply fan and reconcile any differences. For instance, some systems, such as variable air volume systems, allow for a diversity factor. Depending on the size and complexity of the project and systems, general information sheets, instrument calibration sheets, Pitot tube traverse sheets and summary sheets may also be needed. Make a schematic drawing of each system. Schematics may be very detailed showing the central system with pressure and temperature drops across components such as filters, coils, fans, duct sizes, terminal devices, outlets, required and actual flow quantities, etc. At the very least, the schematic should show the general location of the numbered outlets and the location of volume dampers.

Checklist

Before leaving for the site inspection, make a checklist of items needed for the inspection and balance, such as:

1. Data sheets
2. Drawings
3. Tools, ladders, extension cords, flashlight
4. Name of job contacts (engineer, general and mechanical contractors, etc.)
5. Instruments

Evaluate and make plans for the inspection or arrangement of any of the following conditions:

1. Devices, controls or system features which might shut down the system or contribute to an unbalanced condition.
2. General accessibility.
3. Sequence of balancing systems and means of generating an out-of-season heating or cooling load.
4. Marking the final settings of balancing dampers.
5. Arrangements for witnessing of the balancing.

PRELIMINARY FIELD INSPECTION

Walk The System

1. Verify that it is intact and sealed as applicable.
2. It should be clean (no trash, loose insulation or excessive quantities of dirt).
3. Look for missing components such as end caps, dampers, turning vanes, terminal boxes, diffusers, grilles, etc.
4. Ensure that the correct air distribution device is installed. It's not unusual for terminal boxes to be switched.
5. Check that all fire and smoke dampers are installed.
6. Confirm that return air openings are installed for ceiling return plenum systems.
7. Note any major changes in actual installation as compared to design. Make corrections on the schematics. Also, on the schematic make corrections for location changes of dampers, diffusers and other accessories.
8. Inspect the unit equipment.
   a. Motors wired.
   b. Fans and drives mechanically sound and free from debris.
   c. Filters and coils intact and clean.

MOTOR NAMEPLATE INFORMATION

Getting motor nameplate information may occasionally be more difficult than it sounds. Most motors, except some very small ones, would come from the factory with a nameplate. Sometimes, however, the nameplate will be missing, inaccessible, or otherwise, unreadable. In other instances, the nameplates will be positioned so they'll be difficult to read. In many of these cases, a telescoping mirror will help, but be attentive when using a mirror and reading letters and numbers upside down and backwards. For instance, it can be difficult distinguishing between hp and ph, or 5 and 2.

The motor nameplate information that will generally be required on a test report are: phase(s), voltage, amperage, service factor, speed, manufacturer, and horsepower. Other information that may be found on the nameplate which will be needed occasionally are: model number, serial number, Hertz or frequency (Hz or cps), type of motor (induction, etc.), type of electrical connection (wye, delta), power factor, frame size, temperature rise (within the motor), locked rotor amperage, torque characteristics, internal protection, insulation classification, and the wiring diagram.

Phases

In most cases, the motors used on HVAC equipment will be either single-phase or three-phase alternating current, induction motors.

Voltage and Current

Many motors found on HVAC equipment will be dual voltage and depending on how the motors are wired, will operate from either of the voltages listed. A dual voltage motor will also list dual amperage on the nameplate.

Example 2.1: A three-phase, 25 horsepower, dual voltage motor has the following ratings: 230/460 volts, 68/34 amps, 1.10 service factor. This means that if the motor is wired for 230 volts the full load amps will be 68. If it's wired for 460 volts, the amperage will be 34. Notice that volts and amps are inversely proportional. When the voltage doubles the amperage reduces by one-half.

Service Factor

The service factor is the number by which the horsepower or amperage rating is multiplied to determine the maximum safe load that a motor may be expected to carry continuously at its rated voltage and frequency.

Example 2.2: The service factor for the 25 horsepower motor above would allow the motor to operate safely at 27.5 horsepower (25 x 1.10) and about 75 amps (68 x 1.10) with a supply of 230 volts. However, it's not an acceptable practice to leave a motor operating in the service factor area because under certain circumstances it can damage the windings and shorten the life of the motor.

Example 2.3: If this motor was operating at 75 amps and the voltage to the motor dropped to 220 volts because of a reduction in power from the utility or adding other equipment, etc., the amperage would go to about 78 amps (230/220 x 75). The motor would overheat and possibly damage the windings.

Speed

The rated rpm on the nameplate is the speed that the motor will turn when it's operating at nameplate horsepower. If the motor is operating at something other than rated horsepower, the rpm may vary a small amount but it's the nameplate rpm that's recorded on the report forms and used for drive calculations.

Thermal Overload Protection Data

As part of the preliminary balancing procedure, you should check the installed external thermal overload devices. Generally, overload protection is installed to protect the motor against current 125% greater than rated amps. However, in thermal overload selection, it's important to know the ambient temperature at the starter as compared to the ambient temperature at the motor. Sometimes, these temperatures may vary greatly and a temperature-compensating thermal overload or a magnetic overload device may be needed. Check with the manufacturer or supplier for special cases.

Thermal overload protection devices have a letter and/or number on them which is used for proper sizing. A chart listing thermals and their amperage ratings for the specific starters is usually found inside the motor disconnect cover. Thermals must be matched to starter and the rated full load amps of the motor according to the information on the chart or from the manufacturer. If the installed thermal is too large, the motor may not be adequately protected and could overheat. But, if the thermal is too small the motor may stop repeatedly.

If a new thermal is needed the following information will be required: motor starter size, full load amps, service factor, class of insulation, motor classification, and allowable temperature rise. Then referring to the chart or a table provided by the thermal manufacturer, choose the proper size.

DRIVE INFORMATION

To take drive information, stop the fan and put your personal padlock on the motor disconnect switch so only you have control over starting the fan. Remove the belt guard and read the information off the sheaves and belts and measure shaft sizes and distance between the shafts. Also, measure and record the slide adjustment on the motor frame. The slide is for adjusting belt tension. If a sheave has to be changed and room is available on the frame, you may be able to move the motor forwards or backwards to fit the change in sheave size without having to buy a new belt.

Sheaves

Check the outside of the sheave for a stamped part number which shows the sheave size. Record this information on the equipment test report.

Example 2.4: On the fan sheave you might find the word Browning and 3MVB184Q and on the motor sheave, 3MVP70B84P. Referring to the manufacturer's catalog (Browning, in this example) you'd find that the fan sheave has three fixed grooves and can accommodate either a "B" belt at a pitch diameter of 18.4" or an "A" belt at a pitch diameter of 18.0". The outside diameter is 18.75" and the bushing size range is Q1. The letters and numbers on the motor sheave show that it's a three-groove variable pitch sheave with a pitch diameter range of 7.0" to 8.4" for a "B" belt or a range of 6.9" to 8.0" for an "A" belt. The outside diameter is 8.68" and the bushing size range is P2. The bushing sizes Q1 and P2 show that they'll fit shaft sizes from ¾" to $2^{11}/_{16}$" and ¾" to 1¾".

If there's no part number on the sheave, measure the outside diameter and then refer to the manufacturer's catalog to find the corresponding pitch diameter. Most manufacturers list both pitch diameter and outside diameter in their catalogs.

V-Belts

Record the number of belts, manufacturer and size.

Shafts

Measure the size of the motor and fan shafts. Also, measure the distance between the center of the shafts.

START-UP TESTS

Air Distribution

Begin the testing of the system by setting the automatic temperature controls on full cooling (cooling capacity is usually more critical than heating capacity). Check that the cooling coil is dehumidifying and is, therefore, in a "wet" condition. If the coil is "dry" the pressure drop across the coil will be lower than when the coil is wet and therefore, the volume readings will be higher. If the system is balanced in a dry condition, the total airflow will be low when the coil is dehumidifying. However, sometimes it may be necessary to balance with a dry coil. This is all right, as the system will be proportionally balanced; however, the fan speed may need to be increased when the system is rechecked with a wet coil.

Check that automatic temperature controls are properly sequencing and holding the volume dampers in place. If the dampers aren't holding, either because of a malfunction or because the control system isn't finished, disconnect the control linkage and block the dampers in the full cooling position as follows:

1. Systems with face and bypass dampers—the face dampers are full open and the bypass dampers are full closed.

2. Systems with hot and cold deck dampers—the cold deck dampers are full open and the hot deck dampers are full closed. Some systems may have a cooling coil with a diversity factor. Diversity means that the cooling coil is designed for less air than the fan delivers. If this is the case, the total air out of the fan will be divided into the approximate amount required for the cooling coil and the rest will be bypassed.

3. The outside air damper is set approximately at minimum position.

4. The return air damper is full open.

For manual dampers, straighteners and diverters check the following:

1. All smoke and fire dampers are full open, as applicable.

2. All extractors, distribution grids and other accessories are set for maximum air flow.

3. All supply and return dampers including dampers at diffusers and registers are full open.

4. All air pattern devices in diffusers and grilles are properly set.

5. All splitter devices are set in a non-diverting mode.

Fan Rotation

Check the rotation of motors to ensure that fans are rotating in the correct direction. Certain centrifugal fans will produce measurable pressures and some fluid flow, sometimes as much as 50% of design, even when the rotation is incorrect. In axial fans, if the motor rotation is incorrect, the airflow will reverse direction. To check rotation, momentarily start and stop the fan motor to "bump" the fan just enough to determine the direction of rotation. There's usually an arrow on the fan housing showing proper rotation. However, if there's no arrow, view double inlet centrifugal fans from the drive side and single inlet fans from the side opposite the inlet. This will let you determine proper rotation and whether the wheel is turning clockwise or counterclockwise. If the rotation is incorrect, it can be changed in the field. To reverse the rotation on a three-phase motor, change any two of the three power leads at the motor control center or disconnect. In some cases, you may also be able to change rotation in single-phase motors by switching the internal motor leads within the terminal box. Wiring diagrams for single-phase motors are usually found on the motor or inside the motor terminal box.

Observe the fan rotating in the proper direction. If any excessive noises or vibrations are observed, stop the fan and investigate. Before continuing the testing phase, determine if the fan can be operated or if it needs repairing.

Voltage

Voltage measurements are made using a voltmeter which measures the difference in potential between phases, or between phase and neutral. The most accurate voltage will be read at the motor terminal box; however, it's usually much safer to take readings at the motor control center or at the disconnect box. The voltage difference between the two places is generally insignificant. The measured voltage should be plus or minus ten percent of the motor nameplate voltage. If it's not within this range, notify the electrical contractor or the utility company.

On a three-phase motor, if the voltage isn't identical from phase-to-phase, and this is generally the case, there's a voltage unbalance.

When there's a phase voltage unbalance there's also a current unbalance which can be as much as 10 times the percent of voltage unbalance. This means that the motor runs hotter than design which, if the unbalance is large enough, can reduce the life of the motor. Therefore, the maximum allowable phase voltage unbalance for a three-phase motor is 2%. To check for voltage unbalance, use the equation:

% voltage unbalance = maximum deviation from average voltage times 100 divided by average voltage.

Example 2.5: Find the percent of voltage unbalance on a 230 volt, three-phase motor, if the phase-to-phase voltage readings are:

1. T1 − T2 = 242 V
2. T1 − T3 = 239 V
3. T2 − T3 = 227 V

Solving for % of voltage unbalance:

a. Find the average voltage: (242 + 239 + 227)/3 = 236 V

b. Find the maximum deviation from the avg. voltage:

1. = 6V   (242 V − 236 avg. V)
2. = 3V   (239 V − 236 avg. V)
3. = 9V   (236 avg. V − 227 V)

% of voltage unbalance = max. deviation from avg. voltage x 100/avg. voltage

% of voltage unbalance = 9 x 100/236

% of voltage unbalance = 3.8

Since the calculated voltage unbalance exceeds 2%, the electrical contractor or the electric utility company should be notified of the unbalanced condition.

Current

Current is measured using a clamp-on ammeter. The amperage measured on any phase shouldn't exceed the motor nameplate amperage. However, if the operating amperage reading is over the nameplate amperage, take one of the following steps to correct the problem.

1. The reading is over the nameplate amperage but within the

service factor and voltage limits: Reduce the rpm of the fan or close the main discharge damper until the amperage reading is down to nameplate or below.

2. The reading is over the nameplate amperage and outside the service factor limit: Immediately turn the fan off and inform the person responsible for the fan. An exception is if the fan has been running and it's serving a critical area such as a clean room or surgical room, leave the fan operating and immediately notify the person responsible for that fan's operation.

Power Factor

Measuring power factor isn't presently a standard requirement in the testing and balancing industry; however, I've included it in this text for use in determining brake horsepower. Power factor is read with a power factor meter.

Fan and Motor Speed

Two types of instruments, contact and non-contact tachometers, are used for measuring rotational speeds. Readings should be taken until you have two consecutive, repeatable values.

PRESSURE MEASUREMENTS AT THE UNIT

Test Holes

Drill test holes in the fan unit to take air static pressure readings. The holes should be 3/8″ to accommodate the standard Pitot tube. Place the test holes:

1. Before and after the filter for static pressure drop across the filter.

2. Before and after the coil(s), for static pressure drop across the coil(s).

3. Before and after the fan, for total static pressure rise across the fan.

These static pressure readings will be taken using a Pitot tube and a differential pressure gage such as a vertical-inclined manometer.

Reading the Inclined-Vertical Manometer (Fig. 2.1)

First, open both the left and right tubing connectors on the manometer about 1½ turns (both connectors stay open when taking any

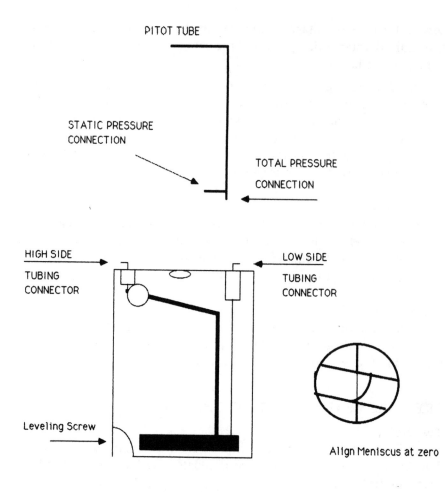

INCLINED-VERTICAL MANOMETER

Figure 2.1

reading; static, velocity or total pressure). Allow enough time for the oil in the manometer to reach ambient temperature. Next, level the manometer using the leveling screw. Align the meniscus to read zero using the adjustment knob on inclined-vertical manometers or the adjustable sliding scale on inclined manometers. You may have to add or remove oil from the manometer to zero it. Careful! It usually takes only a very small amount of oil to fill the manometer. Add a little oil at the tubing connector and let it settle before adding more. Removing the oil is messy, so try to avoid overfilling.

After zeroing the meniscus, check that the manometer is still level. If it is, you're ready to make readings. Ensure that your line of sight is perpendicular to the scale to avoid a false reading (parallax). To help avoid parallax most inclined manometers have a polished, chrome-plated scale. If this is the case, the meniscus must be aligned with its image reflected in the mirrored scale. Only one person should read the manometer through the entire set of readings at any given test hole. Generally, balancing technicians are taught to read the bottom of the meniscus, but reading the top is all right also. Consistency is the key.

Static Pressures Across Filters (Fig. 2.2)

To measure the static pressure drop across the filter, take one piece of the tubing that comes with the manometer kit and connect one end to the static pressure connection on the Pitot tube. The other end of the tubing is connected to the right tubing connector on the manometer. The right connector of the manometer is called the "low" side. The Pitot tube is then inserted in either of the 3/8" test holes drilled in the wall of the unit on both sides of the filter and the static pressure is read. Read the other side of the filter in the same manner. Subtract the two readings.

Static Pressures Across Coils (Fig. 2.2)

If the coils are draw-thru, i.e., before the fan, the static pressure readings are taken in the same manner as the filter pressures. If the coils are blow-thru, i.e., after the fan, the static pressure drop is taken as follows.

1. Connect a piece of tubing from the static pressure port on the Pitot tube to the left side of the manometer (the "high" side).

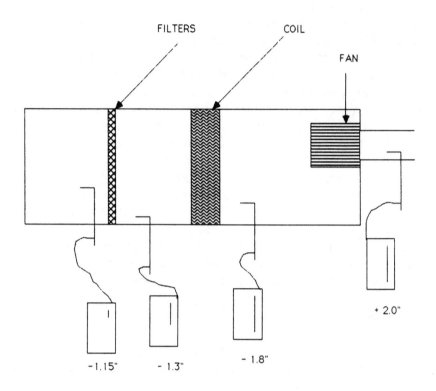

1. FILTER PD = .15 in. wg

2. COIL PD = .50 in. wg

3. TSP = 3.8 in. wg

Figure 2.2

2. Insert the Pitot tube into one of the test holes and take the reading.

3. Change the Pitot tube to the other hole and take the reading.

4. Subtract the readings to get the static pressure drop across the coils.

Static Pressures Across The Fan (Fig. 2.2)

A procedure for measuring total static pressure (TSP) across the fan is to:

1. Connect a piece of tubing from the static pressure port on the Pitot tube to the left side of the manometer ("high" side).

2. Insert the Pitot tube in the test hole on the discharge of the fan and take the static pressure.

3. Switch the tubing from the left side of the manometer to the right side ("low" side).

4. Insert the Pitot tube in the test hole on the inlet of the fan and take the static pressure.

5. Add the readings to get the total static pressure rise across the fan.

To recap:

a. Static pressures taken across components (filters, coils, sound attenuators, dampers, etc.) on the inlet (return side) of the fan will use the right side of the manometer.

b. Static pressures taken across components on the discharge (supply side) of the fan will use the left side of the manometer.

c. Static pressures taken across the fan will use the right side of the manometer for inlet pressures and the left side of the manometer for discharge pressures.

d. If two Pitot tubes are used simultaneously to take static pressures across components, then the Pitot tube sensing the highest pressure, relative to atmospheric pressure, is connected to the high side (left side) of the manometer and the Pitot tube sensing the lower pressure is connected to the low side of the manometer.

Fan Pressures (Figs. 2.2, 2.3 and 2.4)

Fan pressures (fan static pressure and fan total pressure) are used with the fan performance curve to evaluate if the fan is operating as designed.

1. Total pressure (at the fan or in the duct) is the sum of static pressure and velocity pressure (TP = SP + VP and therefore, SP = TP − VP).

2. Fan total pressure is the difference between the total pressure at the fan outlet and the total pressure at the fan inlet (FTP = outlet TP − inlet TP).

3. Fan static pressure is equal to the fan total pressure less the fan velocity pressure (FSP = FTP − FVP).

4. Fan velocity pressure (FVP) is the pressure corresponding to the fan outlet velocity. It's calcualted by dividing the fan air volume by the fan outlet area. Outlet velocity is in reality a theoretical value. It's the velocity that would exist in the fan outlet if the velocity was uniform across the outlet area, which it's not.

To field measure fan static pressure directly, read the fan outlet static pressure and subtract from it the measured fan inlet total pressure (field measured FSP = outlet SP − inlet TP).

	FSP = FTP − FVP
and:	FTP = outlet TP − inlet TP
and:	FVP = outlet VP
therefore:	FSP = outlet TP − inlet TP − outlet VP
and since:	TP = SP + VP
therefore:	outlet TP = outlet SP + outlet VP
therefore:	FSP = outlet SP + outlet VP − inlet TP − outlet VP
canceling:	outlet VP
leaves:	FSP = outlet SP − inlet TP

Example 2.6: The readings on a fan with inlet and discharge ducts are:
  inlet TP = −0.85 in wg (below atmosphere)
  outlet SP = + 1.75" wg (above atmosphere)
      FSP = 1.75 − (−0.85)
      FSP = 2.6" wg

Now that you know how to take fan static pressure, I'm going to

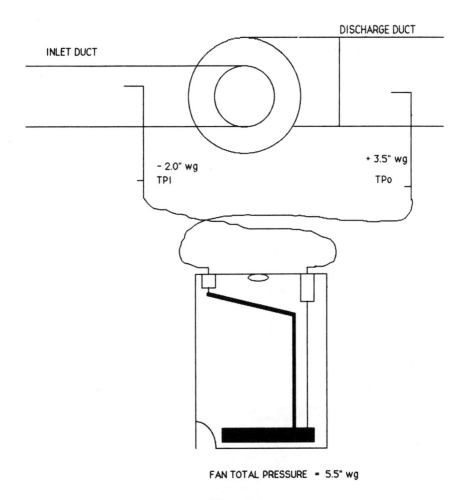

INLET DUCT

DISCHARGE DUCT

- 2.0" wg

TPi

+ 3.5" wg

TPo

FAN TOTAL PRESSURE = 5.5" wg

Figure 2.3

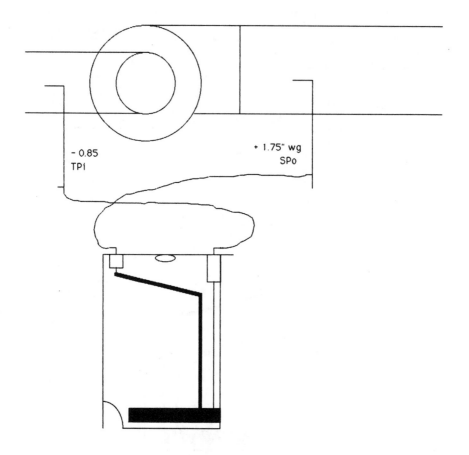

FAN STATIC PRESSURE = 2.6" wg

Figure 2.4

tell you not to try to measure it directly in the field if you're going to apply it to the fan performance curve to determine the operating airflow conditions of the system.

Centrifugal fans are tested and then rated by fan static pressure and in theory, you should be able to measure fan static pressure and fan speed and using the fan performance curve determine airflow quantity. However, fans are almost never installed in a way that will duplicate the conditions under which they were tested and although the field-measured fan static pressure can be plotted on the curve, it won't produce a correct cfm. If, however, a fan static pressure is required, do one of the following:

1. Directly field measure fan static pressure as outlined above.
   a. Note on the report forms that it's field-measured and is probably not a reliable value to be entered on the fan performance curve.
2. Indirectly determine fan static pressure.
   a. Take fan speed.
   b. Get the fan performance curve from the fan manufacturer, if possible, with the field-measured rpm plotted on the curve. If this rpm isn't plotted, you must do it.
   c. Accurately determine total airflow.
   d. Enter the fan curve at the total airflow. Draw a line vertically until it intersects the field-measured rpm curve. At this point draw a line to the left to intersect fan static pressure. This is a reliable fan static pressure which takes into account the fan's system effect.

Instead of directly measuring fan static pressure which is normally not a reliable value, take total static pressure (TSP = outlet SP − inlet SP) across the fan.

Example 2.7: The readings on a plenum supply fan with a discharge duct are:

inlet SP   = −1.8 in. wg (below atmosphere)
outlet SP = + 2.0 inches wg (above atmosphere)
TSP   = outlet SP − inlet SP
TSP   = 2.0 − (−1.8)
TSP   = 3.8″ wg

Since the fan is rated by fan static pressure, total static pressure can't be used in determining airflow capacity from a fan performance curve. Trying to do so will give incorrect results. It can, however, be used to compare the installed duct system static pressures, including pressure losses through filters, coils, etc., to design. It can also be used as a reference of what the fan is doing under the conditions when tested and is considered a more useful measurement for test and balance purposes.

One last word about fan pressures. If the fan is rated by total pressure (such as some axial fans) and a fan total pressure is required, you can measure and add the total pressure on both sides of the fan. However, once again, unless there's adequate straight duct on both sides, this is probably not a reliable value to enter on the fan curve to determine airflow. A better way to determine actual fan total pressure is to measure fan speed and total cfm and use the fan curve as described above.

AIR VOLUME READINGS
IN THE AIR HANDLING UNIT

After starting all the fans (supply, return, and exhaust fans) applicable to the system, the air at the supply unit may be read to give an approximate total cubic feet per minute of supply air volume. The reading will be an approximate value because of the probability of unfavorable field conditions existing at the unit. However, the results will serve as a general check of system performance to make adjustments to the fan speed to bring the system into the proper range for balancing.

Example 2.7: A fan is tested and is 20% over on air. The fan speed is adjusted to bring the air to approximately 10% over on air. Another fan is tested and is 20% low on air. After further investigation, and calculation of horsepower requirements to bring the unit to 100% capacity, it's determined that a new motor will be needed. A decision will have to be made to determine if (1) the system will be balanced at approximately 80% of design and a new motor installed later, (2) a new motor will be installed before continuing the balancing, or (3) the system will be balanced with the air available.

To determine the approximate total air volume in the unit take at least one of the following measurements:

    a. Anemometer traverse across the coil(s).
    b. Anemometer traverse across the filter bank.
    c. Pitot tube traverse in the main supply duct(s).
    d. Pitot tube traverse of the main return duct(s).

Pitot tube traverse of ducts will be discussed in Chapter 3. The anemometer traverse of filters and coils is presented in the next section.

Traverse of Filter Banks and Coils

Where banks of filters or coils are large enough to make them accessible, airflow can be measured by making a traverse with an anemometer. This is not a recommended practice because of (1) the close proximity of the filters to the outside and return air openings and the turbulence of the mixed air and (2) the vena contracta (jet velocities) effect as the air passes through the coils. Any reading will only be an approximation of the total flow. Only experience will tell you if the readings are good or not. The most common instrument used is the 4" rotating vane anemometer, but a capture hood, deflecting vane anemometer, hot-wire anemometer or other electronic air-sensing instrument could also be used. All anemometers are position sensitive so care must be taken when using them. As with all instruments, follow the manufacturer's instructions for best results.

The airflow measurements should be made on the downstream (leaving air) side of the filter or coil. First, measure the filter or coil and calculate square footage. If the filter or coil is very large, divide it into more manageable sections to ease reading. Measure the velocity in each section and determine the average velocity in feet per minute. Multiply the area (in square feet) times the average velocity of each section to find volume in cubic feet per minute (CFM = A x V). Generally, it's been found that this number will be about 25 to 35% high. Therefore, my suggestion is to use this equation: CFM = A x V x 0.70. Add the section volumes together to find total air.

Example 2.8: A coil is 72" x 45" (two sections of 36" x 45"). The average velocity in the sections are: section one - 496 fpm, section two - 585 fpm. The area of each is 11.25 square feet (36 x 45/144).

The cfm for each section is:

One: 3906 cfm (11.25 x 496 x 0.70)
Two: 4607 cfm (11.25 x 585 x 0.70)
Total: 8513 cfm

Another method is using the coil as an orifice and taking the static pressure drop across the coil. Air volume is determined from the manufacturer's chart of "coil cfm vs. coil pressure drop." This isn't a recommended practice because of the difficulty of getting accurate field static pressures and getting reliable rated pressure drop data.

If the pressure measurement is made with a dry cooling coil, the air must be rechecked when the coil is wet as the resistance to airflow will be greater and airflow will be less. Also, if the fan speed is increased based on measurements made on a dry coil (to get desired airflow at the wet coil conditions) recheck the motor current in both the wet and dry conditions to ensure that the motor isn't overloaded.

Chapter 3
Basic Air Balance Procedures
Common To All Systems

This chapter outlines the basic procedures to begin the balance of any air system including how to Pitot tube traverse round, rectangular and flat oval duct, calculate air volume, make air density corrections, and totaling of the system.

BALANCING PRINCIPLES

Balancing is measuring air volumes and adjusting volume control devices to get desired airflow. Fan speeds may also need adjusting. Unless otherwise specified it's generally considered there's an adequate balance when the air quantities measured on the job are within plus or minus 10 percent of desired quantities. The first step in balancing the distribution system is to determine the total air volume. This is accomplished by Pitot tube traverse of main and branch ducts as applicable and the reading of all the supply air outlets.

PITOT TUBE TRAVERSE

If the velocity of the air stream in a duct were uniform, only one reading at any point in the duct would be enough to determine volume of flow. But, this isn't the case. Generally, the velocity, because of friction, is lowest near the sides of the duct, and greatest at or near the center. Therefore, a Pitot tube traverse of the duct is needed to determine the average velocity in the duct at the point of traverse. Having found the average velocity, the volume of air in the duct can then be mathematically calculated using the equation $Q = AV$. Where air quality (Q) in cubic feet per minute is equal to the area (A) of the duct in square feet times the average velocity (V) in feet per minute.

Traverse Location (Fig. 3.1)

Accurate pressure readings can't be taken in a turbulent air stream; therefore, the traverse should be at least 8 duct diameters downstream and 2 duct diameters upstream from elbows, transitions, takeoffs, dampers or other obstructions which cause turbulence. To determine the equivalent duct diameter of a rectangular duct use equation 3.1.

Equation 3.1

$$d = \sqrt{\frac{4ab}{\pi}}$$

where: d = equivalent duct diameter in inches
a = length of one side of rectangular duct in inches
b = length of adjacent side of rectangular duct in inches
π = 3.14

Example 3.1: Find the equivalent duct diameter of a duct that's 24″ wide by 20″ high.

Solution:

$$d = \sqrt{\frac{4ab}{\pi}}$$

$$d = \sqrt{\frac{4 \times 24 \times 20}{3.14}}$$

$$d = 24.7$$

Therefore, a Pitot traverse will be approximately 198″ (8 x 24.7) downstream and 49″ (2 x 24.7) upstream from obstructions.

Field conditions may be such that good locations for Pitot tube traverses can't be found. To determine if the traverse location is good, take a quick set of velocity pressure readings across the duct. If the readings are uniform, the traverse location is probably good. Large variations in the readings show there's considerable turbulence in the duct and not a proper location for the traverse.

Round Duct Traverse — Larger than 10″ (Fig. 3.2)

After the location for the traverse has been determined, drill two

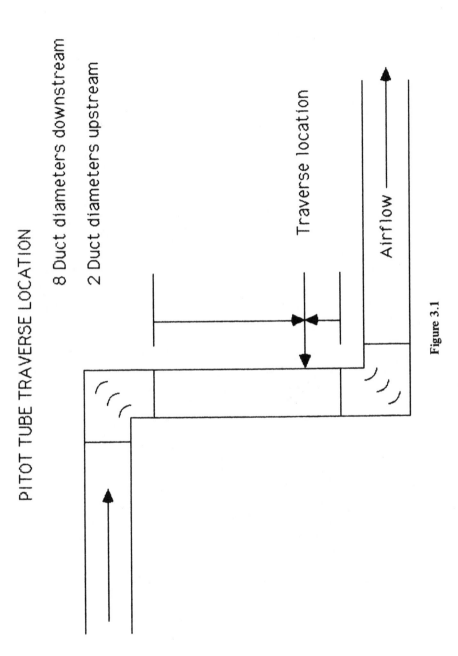

PITOT TUBE TRAVERSE LOCATION

8 Duct diameters downstream

2 Duct diameters upstream

Traverse location

Airflow

Figure 3.1

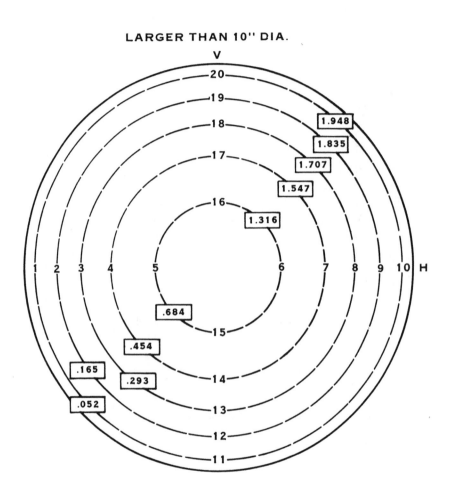

Figure 3.2

holes in the duct 90 degrees apart. Mark the Pitot tube with tape or a marking pen so that 20 velocity pressure readings (10 for each hole) can be taken at centers of equal concentric areas. Pitot tube locations for traversing round ducts are given in fig. 3.2.

Example 3.2: A duct is 20" in diameter. To determine the markings on the Pitot tube, first divide the diameter of the duct by 2 to get the radius. Next, multiply the constant for each traverse point times the radius. The first mark on the Pitot tube, starting from the center of the impact tube would be ½" (to the nearest 1/8"). The fifth mark would be 6-7/8", the sixth mark would be 13-1/8" and the tenth mark would be 19½", etc. For convenience in taking a traverse, the Pitot tube is marked with a stamped number at the even-inch points and odd-inch points are indicated by a 1/8"-long stamped line.

Solution: 20" diameter divided by 2 = 10" radius
1st mark   0.052 x 10" radius = 0.52" = ½"
5th mark   0.684 x 10" radius = 6.84" = 6-7/8"
6th mark   1.316 x 10" radius = 13.16" = 13-1/8"
10th mark   1.949 x 10" radius = 19.49" = 19½"

To check the rounding off calculations, add the 1st and 10th numbers, 2nd and 9th, 3rd and 8th, 4th and 7th, and 5th and 6th. The sum of each will be the diameter of the duct.

Round Duct Traverse — 10" or Smaller (Fig. 3.3)
After the location for the traverse has been determined, drill two holes in the duct 90 degrees apart. Mark the Pitot tube with tape or a marking pen so that 12 velocity pressure readings (6 for each hole) can be taken at centers of equal concentric areas. Pitot tube locations for traversing round ducts are given in figure 3.3. Mark the Pitot tube as described above using the constants for 10" or smaller duct.

Square or Rectangular Duct Traverse (Fig. 3.4)
To make a Pitot tube traverse of a square or rectangular duct determine: (1) the number and distance of the holes in the duct and (2) the markings on the Pitot tube. The number of holes drilled, their distance apart and the number of Pitot tube marks are based on a

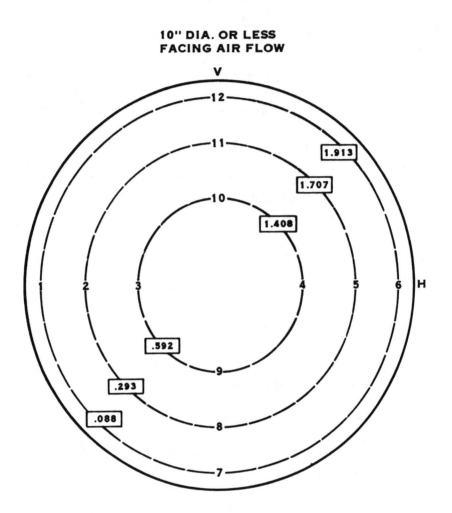

Figure 3.3

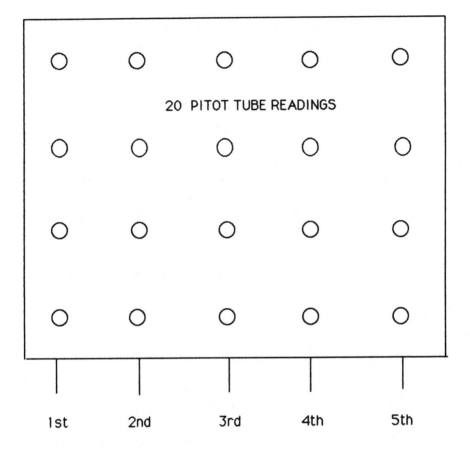

20 PITOT TUBE READINGS

1st   2nd   3rd   4th   5th

TRAVERSE HOLES

Figure 3.4

minimum of 16 velocity pressure readings taken at centers of equal areas and the centers of the areas not more than 6" apart, so that no area exceeds 36 square inches.

Example: 3.3: A duct is 28" x 22". The holes will be drilled in the 28" side. Determine the location of the holes for the traverse.

Solution: Divide 28" by 5. The result is 5 holes drilled at 5.6" apart. Next, divide the 5.6" by 2 to find the first hole from the side of duct (2.8"). The second hole will be 5.6" from the first. The third hole will be 5.6" from the second, etc., as in figure 3.4. If four holes had been chosen, the distance between traverse points would be 7" which is greater than the 6" allowed. If six or more holes had been chosen, it would take more time to drill the extra holes and take the additional readings.

The marks on the Pitot tube are found in the same manner as are the location of the holes.

Example 3.4: Determine the Pitot tube markings for the 28" x 22" duct.

Solution: (1) Divide 22" by 4. $\dfrac{22"}{4} = 5.5"$ on center

(2) Divide the 5.5" by 2 to find the first mark (from the center of the impact tube, 2.75"). The second mark will be 5.5" from the first, etc.

The total readings for a 28" x 22" duct will be 20 (5 holes x 4 Pitot tube points per hole). This meets the criteria of at least 16 readings. Caution: check to see if the duct is lined. The 28" x 22" duct lined with 1" of insulation will have an inside dimension of 26" x 20". The hole placement and Pitot tube markings would change. For example, the holes would be 5.2" apart and the first hole would be 3.6" from the side of the duct (5.2" divided by 2 + 1" for lining).

Flat Oval Duct Traverse (Fig. 3.5)

For a traverse of a flat oval duct, divide the duct into a rectangular area and two semicircles and take a 5-point traverse in each of the semicircles. Next, locate a traverse test hole at the left- and right-hand sides of the rectangle. Then using the criteria of not greater

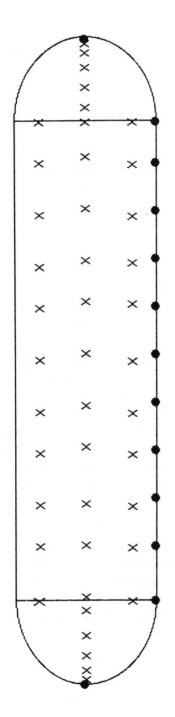

● Location of traverse holes

× Points of measurement

Figure 3.5

than 6" on center, and at least 16 readings, start at the left or right side and find the traverse points for the rest of the rectangle.

ALTERNATE TRAVERSE METHODS

Round Duct Traverse (Log Linear, Fig. 3.6)

After the location for the traverse has been determined, drill three holes in the duct 60 degrees apart. Mark the Pitot tube with tape or a marking pen so that 18, 24 or 30 velocity pressure readings can be taken.

Square or Rectangular
Duct Traverse (Log Tchebycheff, Fig. 3.7)

This method of traversing will require five, six or seven holes drilled into one side of the duct. A recommended procedure is for duct dimensions of 30" or less to have five holes, while duct to 36" will have six holes and seven holes will be drilled in duct greater than 36". The Pitot tube will be marked in a similar manner bringing the total readings to either 25, 30, 35, 36, 42 or 49.

Example 3.5: A 24" x 20" duct would have 25 readings. A 34" x 20" duct would have a 30-point traverse, a 40" x 20" duct would have a 35-point traverse and a 40" x 40" duct would have a 49-point traverse.

Flat Oval Duct Traverse (Fig. 3.8)

After the location for the traverse has been determined, drill two holes in the duct 90 degrees apart. One hole will be for vertical readings and one for horizontal readings. Mark the Pitot tube with tape or a marking pen so that 20 or 40 velocity pressure readings (10 or 20 for each hole) can be taken at centers of equal concentric areas.

Example 3.6: A 48" x 24" flat oval would have the following 20 traverse points:

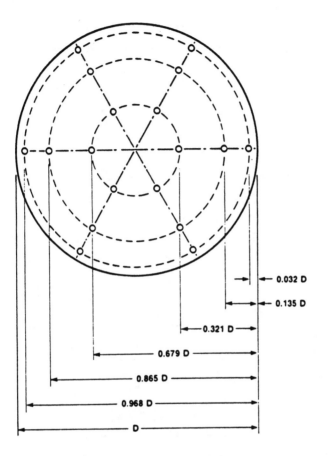

NO. OF MEASURING POINTS PER DIAMETER	POSITION RELATIVE TO INNER WALL
6	0.032, 0.135, 0.321, 0.679, 0.865, 0.968
8	0.021, 0.117, 0.184, 0.345, 0.655, 0.816, 0.883, 0.981
10	0.019, 0.077, 0.153, 0.217, 0.361, 0.639, 0.783, 0.847, 0.923, 0.981

Figure 3.6

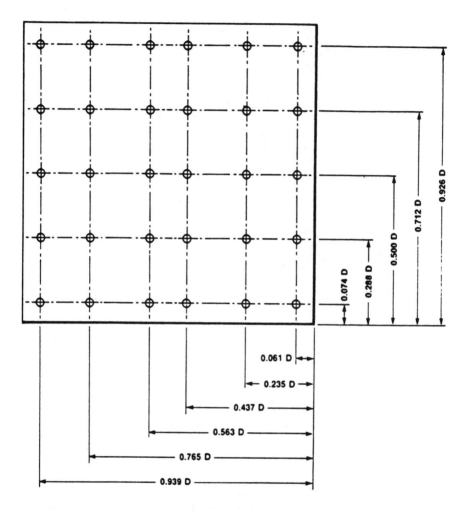

NO. OF POINTS OR TRAVERSE LINES	POSITION RELATIVE TO INNER WALL
5	0.074, 0.238, 0.500, 0.712, 0.926
6	0.061, 0.235, 0.437, 0.563 0.765, 0.939
7	0.053, 0.203, 0.366, 0.500, 0.634, 0.797, 0.947

LOG TCHEBYCHEFF RULE FOR RECTANGULAR DUCTS

Figure 3.7

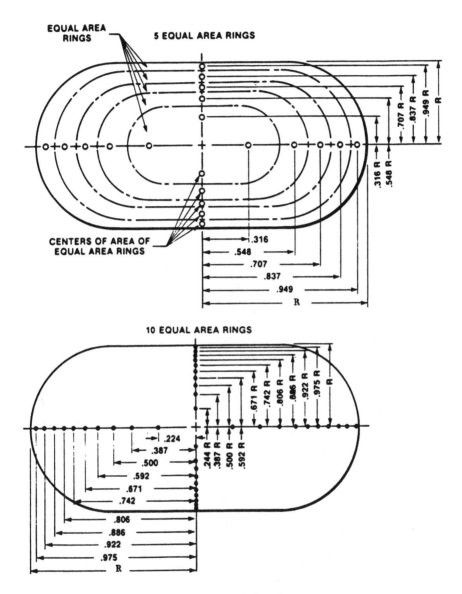

Figure 3.8

Vertical Top to Bottom	Horizontal Left to Right
23.38″	46.75″
22.02″	44.04″
20.48″	40.97″
18.58″	37.15″
15.79″	31.58″
8.21″	16.42″
5.42″	10.85″
3.52″	7.03″
1.98″	3.96″
0.62″	1.25″

Single Point Reading

If the circumstances won't allow a good traverse location, the Pitot tube can be centered in the duct and the centerline velocity pressure can be determined. Convert the velocity pressure to velocity and multiply it by a factor of 0.90 for the approximate average velocity. If you have some straight duct this method will probably be accurate to plus or minus 10%. This type of duct traverse can also be used at good locations for a quick estimation of the average velocity. At good locations the accuracy rises to approximately plus or minus 5%.

VELOCITY PRESSURE READINGS (Fig. 3.9)

To take velocity pressure reading with the manometer, open both the high and low tubing connectors to atmosphere. Using two pieces of tubing connect one piece of tubing to the total pressure connection on the Pitot tube and the other piece of tubing to the static pressure connection. Then connect the total pressure tubing to the left side of the manometer and connect the static pressure tubing to the right side of the manometer. Velocity pressure is the subtraction of static pressure from total pressure and because total pressure is always greater than, or equal to, static pressure, then velocity pressure will always be a positive value. This means that when measuring velocity pressure using the Pitot tube, the hook-up is always the same no matter if the reading is taken on the discharge

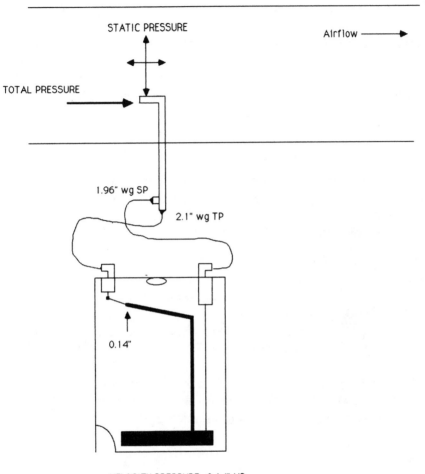

STATIC PRESSURE

Airflow ⟶

TOTAL PRESSURE

1.96" wg SP

2.1" wg TP

0.14"

VELOCITY PRESSURE   0.14" VP

Figure 3.9

of the fan or on the inlet. Check the tubing, especially at the connection ends, for leaks. Also, check that the impact and static holes aren't plugged (particularly when using in insulated ducts). Insert the Pitot tube into the duct facing into the air stream and record the velocity pressure (VP) readings. Continually check to see that (1) the manometer is level, (2) the meniscus is zeroed, and (3) that the Pitot tube is parallel to the air stream.

CALCULATING AIR VOLUME

A Pitot tube traverse is made to determine the average velocity of the airflow in the duct. Since the readings from the manometer are velocity pressure (VP) and not velocity (V), it's necessary to convert VP to V using equation 3.2:

$$V = 4005 \sqrt{VP}$$

Where:

V = velocity in feet per minute (fpm)

4005 = constant

$\sqrt{VP}$ = square root of the velocity pressure in inches of water gage

To get average velocity convert each velocity pressure reading to velocity, total the velocities and divide by the number of readings. Air volume is calculated by using equation 3.3:

$$Q = AV$$

Where:

Q = quantity of air in cubic feet per minute (cfm)

A = cross sectional area of the duct in square feet (sf)

V = velocity in feet per minute (fpm)

Example 3.7: Find the cubic feet per minute flowing in a duct that is 28" x 22" and has an average velocity of 1350 fpm.

Solution: Q = AV

A = 28" x 22"/144 sq. in. per sq. ft.

A = 4.28 sf

Q = 4.28 x 1350

Q = 5778 cubic feet per minute

Caution: Remember, velocity is proportional to the square root of the velocity pressure. Therefore, an incorrect answer will result if the velocity pressures are averaged first and then converted to velocity.

DENSITY CORRECTIONS

Air is less dense as temperature and altitude increase. Therefore, if the air in a duct is different from standard conditions (70 F and 0.075 lbs/cf at 29.92 inches Hg barometric pressure), because of temperature and/or altitude, then the density is also changed. Since most pressure-reading instruments and the equation for converting velocity pressure to velocity are based on the density of standard air, a new density must be calculated. From this calculation, a correction factor for velocity can be determined. The sequence to determine air volume with a correction for density is:

1. Take velocity pressure readings.
2. Convert velocity pressure to velocity.
3. Calculate average velocity.
4. Calculate velocity correction factor.
5. Multiply average velocity times correction factor to get corrected velocity.
6. Multiply corrected velocity times area to get quantity of airflow.

Example 3.8: Find the correction factor for velocity for a duct system in a building in Cheyenne, Wyoming. Elevation 6,000'. Barometric pressure 24 in. Hg.

Solution: (1) Determine the density using equation 3.4:

$$d = 1.325 \frac{Pb}{T}$$

Where:

d = air density in pounds per cubic foot
1.325 = constant
Pb = barometric pressure in inches of mercury
T = absolute temperature (air temperature in degrees Fahrenheit plus 460)

d = 1.325 Pb/T
d = 1.325 x 24/(70 + 460)

$$d = 1.325 \times 24/(530)$$
$$d = 0.060 \text{ pounds per cubic foot}$$

(2) The correction factor for velocity is found by using equation 3.5:

$$cf = \sqrt{\frac{.075}{d}}$$

Where:

cf   = correction factor
0.075 = density of standard air, in pounds per cubic foot
d   = new calculated density, in pounds per cubic foot

Therefore the correction factor for this example is 1.12.

$$cf = \sqrt{\frac{.075}{d}}$$

$$cf = \sqrt{\frac{.075}{.060}}$$

$$cf = 1.12$$

Example 3.9: The average velocity in the 28″ x 22″ duct in the building in Cheyenne, Wyoming, is 1350 fpm. Find the cubic feet per minute.

Solution: (1) Determine the average velocity corrected for density using equation 3.6:

$$Vc = Vm \times cf$$

Where:

Vc = corrected velocity
Vm = measured velocity
cf = correction factor for new density

$$Vc = Vm \times cf$$
$$Vc = 1350 \times 1.12$$
$$Vc = 1512 \text{ fpm}$$

(2) Find the cubic feet per minute.

$$CFM = A \times V_c$$

Where:

  CFM = quantity of airflow in cubic feet per minute
   A = area in square feet
   Vc = corrected velocity

  CFM = A x Vc
  CFM = 4.28 x 1512
  CFM = 6470

A CEFAPP rule of thumb for calculating the velocity correction factor is:

 (1) +2% correction for each 1,000' above sea level
 (2) + or − 1% correction for each 10 degrees above or below 70 degrees Fahrenheit.

Using the rule of thumb the correction factor for 6,000' would be:

 6,000'/1,000' = 6
   6 x 2% = 12%
     cf = 1.12

If the temperature of the air in the duct were 100 degrees Fahrenheit, then the correction factor would be:

 12% for altitude
 +3% for temperature (+1% for each 10 degrees above 70 degrees)
 For a total of 15% or a correction factor of 1.15.

Checking the rule of thumb:

$$d = 1.325 \frac{Pb}{T}$$

$$d = 1.325 \times 24/(100 + 460)$$

$$d = 0.057$$

$$cf = \sqrt{\frac{.075}{d}}$$

$$cf = \sqrt{\frac{.075}{.057}}$$

$$cf = 1.15$$

A CEFAPP rule of thumb for determining barometric pressure is:

0.1 inch Hg reduction from 30" (29.92 rounded off) for each 100' above sea level.

Example 3.10: 5600' elevation would be 56 x 0.1" = 5.6".
$$30" - 5.6" = 24.4"$$

TOTALING THE SYSTEM

If Pitot tube traverses of the system won't produce total airflow, a complete reading of the supply air outlets or a combination of the two will. Whatever the situation, it's necessary to read the entire system for balancing. Sometimes, it may be necessary to combine a Pitot tube traverse with reading the outlets to determine outlet air quantity. For instance, a branch duct has four outlets on it. Three of the four outlets can be read but the fourth is inaccessible. A Pitot tube traverse is made of the branch and the difference between the total cfm of the three outlets and the cfm from the traverse is the cfm of the fourth outlet.

It's usual for the cfm at the outlets to be less than the cfm at the traverse because of some duct leakage, especially in unsealed low pressure systems. However, if the difference is more than 10%, search for leaks or obstructions in the ductwork. The reason for the low airflow may be determined by a visual check or by taking a static pressure profile of the duct or outlet in question. Make corrections, if possible, for any blockages (such as a closed damper, loose insulation, etc.), loss of air (from a leaking duct or an end cap is off), or excessive friction loss because of poor duct design or installation. Notify the sheet metal installer if it's found that the ductwork is leaking excessively or the installation is poor. Quite often a volume damper handle or damper quadrant will indicate an open damper when the damper may be full or partly closed. If the damper can't be visually checked, take a static pressure drop across it. The pressure drop will be very low if the damper is wide open.

A higher total cfm at the outlets than at the traverse is probably from inaccuracies in reading the traverse and/or outlets.

Problem Areas

If, after the total air has been determined, the system isn't within design tolerances (for most systems, the specifications call for balancing at plus or minus 10% of design) try to determine the reasons for the difference and make changes as needed. For instance, check for:

1. Correct diversity.
2. Excessive pressure drops across system components such as filters, coils, sound attenuators, etc. Consult the manufacturer's data.
3. Poor duct design or duct installation at the fan inlet and outlet.
4. Radical fittings in the duct system.

If a part of the system is extremely low, and the reason for the low airflow can't be corrected before starting the balancing (for instance, because the problem is poor duct design), simply ignore the problem area and balance the rest of the system. Don't sacrifice the entire system balance by reducing airflow in most of the system in an attempt to force air to the problem area. After the balance is finished go back to the problem area and make corrections.

If the balance can be delayed, or corrections need to be made after the balance, consult with the design engineer for solutions. For example (1) remove restrictive ductwork, (2) design new ductwork, (3) remove the duct from the present system and add it to another system, or (4) design and install a new, separate HVAC system to handle the space.

Adjusting Fan Speed

If it's determined that everything is in order, but the total air is still below design, you may want to consider increasing the fan speed to bring the airflow as close to 110% of design as possible.

During the balancing process (i.e., changing the position of dampers) there will be some loss of total air, generally between 5 and 10%. In theory, whenever a damper is closed the static pressure upstream of the damper is increased and since the fan is working against a greater static pressure there will be a decrease in total air. Therefore, when several dampers are throttled, it can be expected there will be an increase in static pressure at the discharge of the supply fan (an increase in static pressure at the inlet of a return or exhaust fan) and a reduction in total air. This is why it's good practice, if possible, to set the fan at 110% of design to allow for the loss of air during dampering. Even with this cushion, it still may be necessary to increase the fan speed when the balancing is finished to bring the airflow to desired conditions. However, it's not absolutely necessary that the speed be increased now; it can be done after the balancing is finished.

If the fan speed is to be increased, be sure to use the fan laws to calculate not only the needed adjustment to the sheaves and belts, but also for the new, required brake horsepower. It's critical when increasing fan speed to determine if there's adequate horsepower available. Never increase the fan speed to a point where the motor is in an overloaded condition. Also, if a sizeable upward change is being made, check the manufacturer's fan rating table to determine the class of fan and the maximum allowable rpm. Contact the fan manufacturer if there's a question about what affect an increase in speed will have on the fan.

If, on the other hand, the airflow is greater than 20% above design (120% of design), you may want to reduce the fan speed to bring the air volume down to 10 to 15% above design.

Chapter 4
Proportionally Balancing
Constant Volume
Low Pressure Systems

This chapter outlines the procedure for proportionally balancing low pressure constant volume systems and the low pressure side of any system whether it be a constant or variable air volume system; low, medium, or high pressure system; or single zone, multizone, or dual duct system.

INSTRUMENTATION

Anemometers

In the field, the airflow through the outlets (or inlets) may be measured using a rotating vane anemometer for sidewalls (grilles and registers), or a deflecting vane anemometer for sidewalls and ceiling diffusers. Anemometers require a correction or flow factor for each outlet to convert velocity readings to cfm. In addition to the manufacturer's flow factor for the outlets/inlets, the anemometer may also have a calibration or correction factor.

Capture Hoods

A capture hood may also be used to measure sidewalls and ceiling diffusers. The capture hood is the easiest and most reliable instrument to take outlet/inlet readings because the airflows are read directly in cfm and flow factors aren't needed. However, if a capture hood reading is in question, and a correction factor seems indicated, because of extraordinarily high or low velocities or an unusual use of the capture hood, take a traverse of the duct and determine a correction factor. In addition to a field-measured correction factor for air devices, check the capture hood for a manufacturer's calibration or correction factor.

Example 4.1: Readings are taken on several 5' linear air diffusers (LAD) located in a drop ceiling. The capture hood only has a 2' x 2' bonnet. It's decided that the procedure will be to read the first two feet, then the second two feet and then the last foot of the diffuser and add the three readings together for a total airflow through the LAD. To determine if a field correction factor will be needed, a Pitot tube traverse will be made and the readings will be compared with the capture hood readings. The cfm calculated at the outlet divided by the cfm at the traverse point is the correction factor. Indications are that some capture hoods may read low when measuring linear air diffusers.

Flow Factors

A flow factor, sometimes called a K-factor or Ak factor is the effective area of the grille or diffuser as determined by the manufacturers' own airflow tests. Check with the manufacturer for a catalog of grilles and diffusers and their corresponding flow factors. These flow factors must be used to calculate cfm ($Q = Ak \times V$). It's also important to understand that these flow factors apply only when using the instrument specified by the outlet manufacturer and in the prescribed method.

If a flow factor is not available or it's producing an unsatisfactory result, a flow factor can be field-determined if it's possible to take a Pitot tube traverse in the supply duct for the outlet in question. The supply duct must be free from obstructions and air leaks from the point of traverse to the outlet. Take readings at the traverse point and at the outlet. The cfm calculated at the outlet divided by the cfm at the traverse point is the flow factor.

PROPORTIONALLY BALANCING LOW PRESSURE SYSTEMS

The principles of proportionally balancing require that all the dampers in the distribution system be full open and that at least one outlet volume damper (the outlet with the lowest percent of design flow) will remain open. If the system has branch ducts, at least one branch volume damper (the branch with the lowest percent of design flow) will also remain full open. Because the air outlets are on the low pressure side of any system the following proportional balancing procedure can be used on constant or variable air volume systems: low, medium, or high pressure systems; or single zone, multizone, or dual duct systems.

Procedure:

1. Determine which outlet has the lowest percent of design flow (%D). Typically the outlet with the lowest %D will be on the branch farthest from the fan. This outlet will be called the "key" outlet.

    a. Design flow is either the original design flow per the contract specifications or a new calculated design flow.

    b. Percent of design flow is equal to the measured flow divided by the design flow:

    $$\%D = \frac{M}{D}$$

    c. If anemometers are being used the measured and design flows will be in feet per minute (fpm), whereas if capture hoods are used, the flows will be in cubic feet per minute (cfm). All the system balancing examples in this book will use cfm as the measured flow.

2. Starting with the key outlet, as needed, adjust each outlet on that branch duct in sequence, from the lowest percent of design flow to the highest percent of design flow.

    a. The ratio of the percent of design flow between each outlet must be plus or minus 10%. Ratio of design flow is equal to outlet X %D divided by outlet Y %D.

    b. To reduce airflow, volume dampers in the system should be adjusted in the branch ducts and at the takeoffs and not at the outlet since dampering at the outlets results in excessive noise and poor air distribution.

3. Go to the branch that has the outlet with the next lowest percent of design flow as determined from the initial readout. Typically, this "key" outlet will be on the second farthest branch. Balance all the outlets on this branch to the key outlet to within plus or minus 10% of design flow.

4. Continue until all the outlets on all the branches have been balanced to within plus or minus 10% of each other.

5. Starting with the branch with the lowest percent of design flow as the key branch proportionately balance all branch ducts from the lowest %D flow to the highest %D flow to within 10% of each other.

6. Continue until all branches have been balanced.

7. Adjust the fan speed if needed to bring the system to within 10% of design flow.

8. Reread all the outlets and make any final adjustments.

Example 4.2: Balance the constant volume low pressure system illustrated in figure 4-1.

Conditions:

1. The system has been inspected and all dampers are open. The fan is operating correctly. Traverse and static pressure readings have been taken at points A, B and C. The results are:

   Point A: 2185 cfm (115%D)
   Point B: 975 cfm (122%D)
   Point C: 1180 cfm (107%D)

2. All the outlets have been read and the results are:

Outlet	Design	Measured
1	250	235
2	300	280
3	250	265
4	300	340
5	200	230
6	200	210
7	200	260
8	200	225
Total	1900	2045

Branch	Design	Measured	%D
B	800	925	116
C	1100	1120	102
Total	1900	2045	

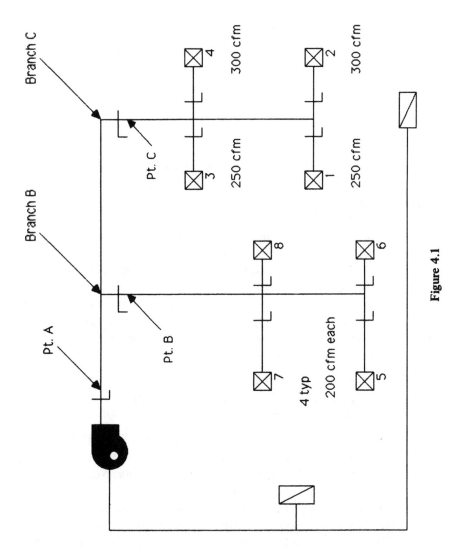

Figure 4.1

Balancing Procedure:

1. Determine which outlet has lowest percent of design flow.

Outlet	Desired	Measured	%D
1	250	235	94
2	300	280	93
3	250	265	106
4	300	340	113
5	200	230	115
6	200	210	105
7	200	260	130
8	200	225	113
Total	1900	2045	108

2. Outlet No. 2 is the key outlet. Therefore, the balance will start with outlet No. 2 on branch C. The balancing damper at the takeoff to outlet No. 2 will remain open.

    a. Outlet No. 1 is the next highest %D at 94%.

    b. Determine if No. 1 and No. 2 are plus or minus 10% of each other. The ratio of No. 1 to No. 2 is 1.01 (94%/93%). The ratio between No. 1 and No. 2 is within 10%. This ratio will remain in effect for as long as the dampers for the two outlets aren't moved. Therefore, whatever percent of design flow outlet No. 2 goes to, outlet No. 1 will be 1.01 times it. For example, if the airflow to outlet No. 2 is brought to 100%D, outlet No. 1 will be 101%D.

Outlet	Design	Measured	%D	Ratio
1	250	235	94	1:2 = 1.01
2	300	280	93	

    c. Go to the outlet with the next highest %D outlet No. 3. Compare outlet No. 3 to outlet No. 2. The ratio of No. 3 to No. 2 is 1.14 (106%/93%). The ratio between No. 3 and No. 2 is greater than 10% (1.10).

    d. To balance No. 3 to No. 2, the volume damper for outlet No. 3 will be closed. Arbitrarily close No. 3 to 100%D or 250 cfm. Go back and read No. 2. It reads 285 cfm. Determine %D for No. 3 and No. 2 and the ratio between them.
    No. 3 is 100%D
    No. 2 is 95% D (285/300)

The ratio between No. 3 and No. 2 is 1.05 (100%/95%)

Since the ratio is within 10%, the damper on No. 3 is locked down. Once the ratio between two outlets is set between plus or minus 10% the damper is locked down and it's not moved again until the final fine tuning.

Outlet	Design	Measured	%D	Ratio
1	250	240*	96*	1:2 = 1.01
2	300	285	95	
3	250	250	100	3:2 = 1.05

*Calculated for the purposes of this exercise only to illustrate what is happening at the outlets already set. Once an outlet or branch damper is set the cfm is normally not reread until the final balance and generally there is no need to calculate the airflow.

    e. Go to outlet No. 4 and read it. In this example it still reads 340 cfm, 113%D. Outlet No. 4 and No. 2 aren't within 10% of each other (113%/95%). Arbitrarily cut No. 4 to 100%D or 300 cfm. Read No. 2. It reads 295 cfm. Determine %D for No. 4 and No. 2 and the ratio between them.

No. 4 is 100%D

No. 2 is 98%D

The ratio between No. 4 and No. 2 is 1.02 (100%/98%).

Since the ratio is within 10%, the damper on No. 4 is locked down. All the outlets on branch C have been balanced.

Outlet	Design	Measured	%D	Ratio
1	250	248*	99*	1:2 = 1.01
2	300	295	98	
3	250	258*	103*	3:2 = 1.05
4	300	300	100	4:2 = 1.02

*Calculated

3. Go to the key outlet on branch B. This will be outlet No. 6 at 105%D. Balance the other outlets to No. 6.

Outlet	Design	Measured	%D
5	200	230	115
6	200	210	105
7	200	260	130
8	200	225	113

a. No. 8 to No. 6 is 1.08 (113%/105%).

Outlet	Design	Measured	%D	Ratio
5				
6	200	210	105	
7				
8	200	225	113	8:6 = 1.08

b. No. 5 to No. 6 is 1.1 (115%/105%)

Outlet	Design	Measured	%D	Ratio
5	200	230	115	5:6 = 1.1
6	200	210	105	
7				
8	200	225	113	8:6 = 1.08

c. No. 7 to No. 6 is 1.24 (130%/105%). The volume damper for
    outlet No. 7 is arbitrarily cut so that outlet No. 7 reads
    115%D or 230 cfm. Read No. 6. It reads 220 cfm. Determine
    %D for No. 7 and No. 6 and the ratio between them.

    No. 7 is 115%D

    No. 6 is 110%D

    The ratio between No. 7 and No. 6 is 1.05 (115%/110%).
    Since the ratio is within 10%, the damper on No. 7 is locked
    down. All the outlets on branch B have been balanced.

Outlet	Design	Measured	%D	Ratio
5	200	242*	121*	5:6 = 1.1
6	200	220	110	
7	200	230	115	7:6 = 1.05
8	200	238*	119*	8:6 = 1.08

*Calculated

Branch Balancing Using Outlets

To proportionally balance the branches, start with the branch
with the lowest percent of design airflow. You can balance branches
using a representative outlet on each branch or using static pressures.

Example 4.3: Balancing the constant volume low pressure system
illustrated in figure 4-1 using representative outlets on each branch.

1. After proportionally balancing the outlets the results are:

Outlet	Design	Measured	%D	Ratio
1	250	248*	99*	1:2 = 1.01
2	300	295	98	
3	250	258*	103*	3:2 = 1.05
4	300	300	100	4:2 = 1.02
5	200	242*	121*	5:6 = 1.10
6	200	220	110	
7	200	230	115	7:6 = 1.05
8	200	238*	119*	8:6 = 1.08

*Calculated

2. Use outlet No. 2 and No. 6 to represent their branches and to determine the branch with the lowest percent of design flow. Outlet No. 6 is 110%D and outlet No. 2 is 98%D. The ratio between the outlets is 1.12 (110%/98%).

   a. The volume damper on branch B is arbitrarily closed to bring outlet No. 6 down to 105%D (210 cfm). Read outlet No. 2. It reads 300 cfm. Determine %D for No. 6 and No. 2 and the ratio between them.

   No. 6 is 105%D
   No. 2 is 100%D

   The ratio between No. 6 and No. 2 is 1.05 (105%/100%)
   Since the ratio is within 10%, the branch volume damper on branch B is locked down. The branches are proportionally balanced to each other. As all the outlets have been proportionally balanced to each other by branch, an adjustment at the branch damper will increase or decrease all the outlets on each branch proportionally.

Outlet	Design	Measured	%D	Ratio
1	250	253*	101*	1:2 = 1.01
2	300	300	100	
3	250	263*	105*	3:2 = 1.05
4	300	306*	102*	4:2 = 1.02
5	200	230*	115*	5:6 = 1.10
6	200	210	105	6:2 = 1.05 (branch B/C)
7	200	220*	110*	7:6 = 1.05
8	200	226*	113*	8:6 = 1.08

*Calculated

3. Reread all the outlets and make any final adjustments. Record final readings on the report forms.

Outlet	Design	Measured	%D
1	250	250	
2	300	300	
3	250	265	
4	300	310	
5	200	225	
6	200	210	
7	200	220	
8	200	220	
Total	1900	2000	105

Branch	Design	Measured	%D
B	800	875	109
C	1100	1125	102
Total	1900	2000	105

4. Since the system is within 10% of total design flow, no further adjustments are needed.

Branch Balancing Using Static Pressure

Another method for proportionally balancing branches is using static pressure. If a traverse of the branch was made, use this as total cfm or use the total of the outlets. Next, take a static pressure reading, if it hasn't already been done, at the traverse point. These readings will be called CFM1 and SP1. Use fan law No. 2 to calculate the required static pressure (SP2) that will result in the design airflow (CFM2).

When using the fan laws it's absolutely necessary to ensure that no adjustments or changes are made downstream of the branch damper during the branch balancing process. To illustrate what would happen let's say that after taking the branch traverse and static pressure a damper is closed in one of the outlet takeoffs. This means that the total cfm at the traverse will be reduced and the static pressure will be increased. Just the opposite would happen if an outlet damper were found closed and then opened after the traverse and static pressure had been taken; the cfm would increase and the static pressure would decrease at the traverse point. Under both these circumstances the fan law couldn't be used until a new cfm and static pressure were taken after the changes to the system had been made. Therefore, it's extremely important to ensure that all the dampers are open before starting the balance and that nothing is disturbed during the balance.

Example 4.4: Balancing the constant volume low pressure system illustrated in figure 4-1 using branch static pressures.

Initial conditions:

Location	Design CFM2	Measured CFM1	%D	SP1	Ratio
Point B	800	930	116	0.70	
Point C	1100	1100	100	0.52	B:C 1.16

1. To bring the branches within 10% of each other it will be necessary to close the volume damper on branch B.
   a. Arbitrarily elect to close branch B to 108%D.
   b. Using fan law No. 2 determine what new static pressure at Point B will correspond to 108%D or 864 cfm.

$$\left(\frac{CFM_2}{CFM_1}\right)^2 = \frac{SP_2}{SP_1}$$

$$SP_2 = SP_1 \left(\frac{CFM_2}{CFM_1}\right)^2$$

$$SP_2 = 0.70 \left(\frac{864}{930}\right)^2$$

$$SP_2 = 0.60 \text{ in. wg}$$

   c. Close the volume damper on branch B until you read 0.60 inches wg SP on the gage. Go to Point C and read the new static pressure. It reads 0.57 inches wg. Calculate new cfm, %D flow and ratio.

$$\left(\frac{CFM_2}{CFM_1}\right)^2 = \frac{SP_2}{SP_1}$$

$$CFM_2 = CFM_1 \sqrt{\frac{SP_2}{SP_1}}$$

$$CFM_2 = 1100 \sqrt{\frac{0.57}{0.52}}$$

$$CFM_2 = 1155$$

Location	Design CFM	Measured CFM	SP	%D	Ratio
Point B	800	864	0.60	108	
Point C	1100	1155	0.57	105	B:C 1.03

d. The branches are now balanced to each other within +/− 10% and since all the outlets have been proportionally balanced to each other by branch, an adjustment at the branch damper will increase or decrease all the outlets on each branch proportionally. The ratio of the outlets remains the same as when initially set since no outlet volume dampers have been changed. To determine the cfm at each outlet, read outlets No. 2 and No. 6 and calculate new cfm for the other outlets.

Outlet	Ratio
1	1:2 = 1.01
2	Measured cfm
3	3:2 = 1.05
4	4:2 = 1.02
5	5:6 = 1.10
6	Measured cfm
7	7:6 = 1.05
8	8:6 = 1.08

Chapter 5
Balancing Constant Volume Multizone, Dual Duct And Induction Systems

MULTIZONE (MZ) SYSTEMS

Multizones are dual path systems usually having a cooling coil and heating coil. The air passes through the coils into mixing dampers and then into zone ducts to the various conditioned spaces. Some systems won't have a heating coil but will instead bypass mixed air around the cooling coil when the system is in a heating mode. If this is the case, after the system is balanced on full cooling set it to full heating and check the cfm through the resistance apparatus in the hot deck. The cfm through the resistance apparatus, which may be a damper or a perforated plate, should be about the same as the airflow through the cooling coil when the system is in the cooling mode.

Multizone systems are designed as constant volume systems but the actual volume may vary to 10% during normal operation because of the changes in resistance between the smaller heating coil and the larger cooling coil.

Balancing Procedure
1. Do the preliminary office work.
   a. Gather plans and specifications.
   b. Prepare report forms.

2. Do the preliminary field inspection.
   a. Inspect the job site.
   b. Inspect the distribution system.
   c. Inspect the equipment. Where applicable check that the unit's mixing dampers are operating correctly with minimal leakage. Depending on circumstances this may be done visually, by reading temperatures, or using static pressure drops.

3. Make initial tests of all the equipment applicable to the system being balanced.
   a. Check rotation.
   b. Take electrical measurements.
   c. Take speed measurements.
   d. Set all dampers at the full open position except the outside air which is set at minimum position.
   e. Operate all fans (supply, return and exhaust) at or near design speeds.
   f. Take static pressure measurements.
   g. Take total air measurements. If the system has diversity, determine the diversity ratio and keep the proportion of cooled air to total volume constant during the balance by (1) setting some mixing dampers so the hot deck dampers are partially open at full cooling or (2) setting enough zones to full cooling to equal the design flow through the cooling coil. The remaining zones will be set to heating.

4. Balance and adjust the distribution system.
   a. Set the zones being tested on full cooling.
   b. Proportionally balance the outlets.
   c. Proportionally balance the branches.
   d. Proportionally balance the zones.

5. Change fan speed as needed.

6. Take final readings in the cooling mode and check the system in the heating mode.

7. Check the system in the maximum outside air mode. If the motor overloads or the flow is excessive, adjust outside air dampers or fan speed as needed.

8. Complete report.

PROPORTIONALLY BALANCING ZONES

The procedure outlined above is the same as earlier discussed for low pressure constant volume systems except for the diversity and proportionally balancing the zones.

Balancing Procedure:

1. After the outlets and branches are proportionally balanced, take

a centerline static pressure reading after the zone volume damper (Fig. 5.1).

2. Using the air volume from a Pitot tube traverse of the zone duct or the total of the zone outlets, calculate the percent of design flow for each zone.

3. Starting with the zone with the lowest percent of design flow, proportionally balance the zones using static pressure readings and fan law No. 2.

Multizone systems generally have between 4 and 12 zones. In most cases, the zones will be similar in flow quantities. If, however, any zones are quite different in flow quantity than the other zones, balance them separately. The sequence of balancing will depend on the circumstances. For example, a system has 7 zones. Zones 1 through 6 are sized for 1200 to 1500 cfm each and they're all within plus or minus 20% of each other. The seventh zone is sized for 500 cfm and has a measured airflow of 115% of design. Here, the sequence will be to proportionally balance zone 1 through 6 and then read zone 7 and damper it.

Example 5.1: Using figure 5.1 proportionally balance the zones within plus or minus 10% of each other.

Conditions:

Zone	Design	Measured	%D	SP1
1	1000	1350	135	1.2
2	1000	1200	120	1.0
3	1000	950	95	.92
4	1000	810	81	.83
5	1000	890	89	.79

1. Determine which zone has the lowest %D. Zone 4.

2. Compare the %D of zone 5 and zone 4. The ratio is 1.10. The zones are proportioned.

3. Determine ratio of zone 3 to zone 4. The ratio is 1.17 (95%/81%).

4. Arbitrarily close zone 3 volume damper until a static pressure of 0.77 is read. This will give zone 3 approximately 87%D.

$$SP_2 = SP_1 \left(\frac{\%D_2}{\%D_1} \right)^2$$

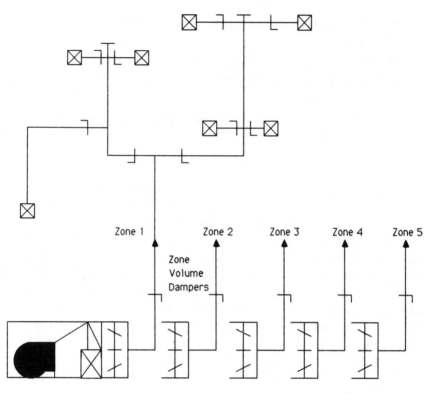

Figure 5.1
MULTIZONE SYSTEM

$$SP_2 = 0.92 \left(\frac{87}{95}\right)^2$$

$$SP_2 = 0.77 \text{ in. wg}$$

5. Read zone 4 static pressure and calculate %D.
    a. Zone 4 reads 0.85 inches wg.
    b. The approximate %D is 82%.

$$\%D_2 = \%D_1 \sqrt{\frac{SP_2}{SP_1}}$$

$$\%D_2 = 81 \sqrt{\frac{.85}{.83}}$$

$$\%D_2 = 82$$

6. Determine new ratio of zone 3 to zone 4.
    a. It's 1.06 (87%/82%).
    b. Zone 3, 4 and 5 are now proportionally balanced.

Zone	Design	Measured	%D	SP	Ratio
1	1000				
2	1000				
3	1000	870	87	.77	3:4 = 1.06
4	1000	820	82	.85	
5	1000	920	90	.81	5:4 = 1.10

7. Balance zone 2 to zone 3 and then balance zone 1 to zone 2.

8. Recheck static pressures. If no further balancing is needed, make final readings of outlets.

DUAL DUCT (DD) SYSTEMS

Dual duct or double duct systems are dual path systems usually having a cooling coil and heating coil. The air passes through the coils into the hot and cold ducts that run to the space where they terminate into mixing boxes which supply heated, cooled or mixed air to the conditioned space. Most dual duct systems are designed as constant volume, medium to high pressure, pressure independent systems. Two other categories of dual duct systems are:

1. Constant volume, low pressure, pressure dependent systems. These systems don't use mixing boxes. They use mixing dampers in the hot and cold duct to supply mixed air to the space through a common secondary duct.

2. Variable air volume, pressure independent systems. This chapter will outline the procedure for balancing constant volume dual duct systems. Balancing VAV dual duct systems will be discussed in Chapter 6.

Balancing Procedure for Medium to High Pressure, Pressure Independent Systems (Fig. 5.2)

1. Do the preliminary office work.
   a. Gather plans and specifications.
   b. Prepare report forms.

2. Do the preliminary field inspection.
   a. Inspect the job site.
   b. Inspect the distribution system. Where applicable check that (1) the duct has been leaked tested and (2) the mixing box dampers are operating correctly with minimal leakage. Depending on circumstances, this may be done visually or by reading temperatures.
   c. Inspect the equipment.

3. Make initial tests of all the equipment applicable to the system being balanced.
   a. Check rotation.
   b. Take electrical measurements.
   c. Take speed measurements.
   d. Set all dampers at the full open position except the outside air which is set at minimum position.
   e. Set the system for full cooling.
   f. Operate all fans (supply, return and exhaust) at or near design speeds.
   g. Take static pressure measurements at the unit.
   h. Take measurements at the end of the system to determine if

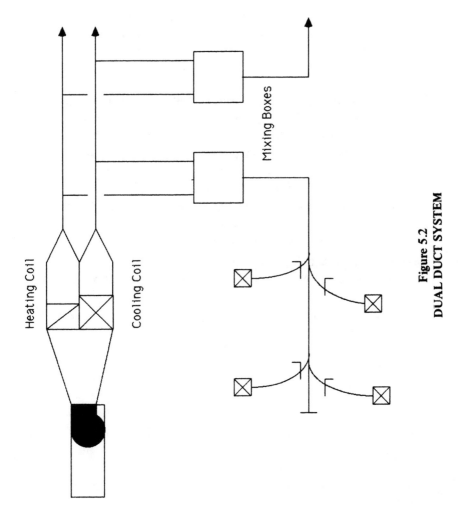

Figure 5.2
DUAL DUCT SYSTEM

the static pressure is at or above the minimum required for mixing box operation. Increase or decrease fan speed as needed. Check the static pressure drop across the box. Consult box manufacturer for pressure drop requirements. Generally, 0.75 inches wg is required for mechanical regulators. Additional pressure is needed for the low pressure distribution system downstream of the box: approximately 0.1 inch wg per 100 foot of duct (equivalent length) and 0.05 to 0.1 inch wg for the outlet.

i.   Take total air measurements. If all the boxes are constant volume, set the thermostats for full flow through the cold duct. Traverse the cold and hot ducts. If more than 10% of the total rated airflow is measured in the hot duct, check for leaking hot air dampers or boxes with crossed supplies. If the system has diversity, determine the diversity ratio and keep the proportion of cooled air to total volume constant during the balance by setting enough boxes to full cooling to equal the design flow through the cooling coil. The remaining boxes will be set to heating.

4. Balance and adjust the distribution system.
   a.   Set the box regulators according to manufacturer's recommendations for airflow through the boxes.
   b.   Take Pitot tube traverses of the low pressure duct off the boxes and/or a total of the outlets to confirm box setting and determine duct leakage.
   c.   Proportionally balance the outlets.

5. Change fan speed as needed.

6. Take final readings and check the system in the heating mode.

7. Check the system in the maximum outside air mode. If the motor overloads or the flow is excessive, adjust outside air dampers or fan speed as needed.

8. Complete report.

Balancing Procedure for Low Pressure,
Pressure Dependent Systems

1. Do the preliminary office work.
   a.   Gather plans and specifications.

    b. Prepare report forms.

2. Do the preliminary field inspection.
    a. Inspect the job site.
    b. Inspect the distribution system.
    c. Inspect the equipment.

3. Make initial tests of all the equipment applicable to the system being balanced.
    a. Check rotation.
    b. Take electrical measurements.
    c. Take speed measurements.
    d. Set all dampers at the full open position except the outside air which is set at minimum position. Verify that automatic static pressure dampers, if installed, are open.
    e. Set the system for full cooling.
    f. Operate all fans (supply, return and exhaust) at or near design speeds.
    g. Take static pressure measurements at the unit.
    h. Take total air measurements. Traverse the cold and hot ducts. If more than 10% of the total rated airflow is measured in the hot duct, check for leaking hot air dampers or crossed supplies.

4. Balance and adjust the distribution system.
    a. Proportionally balance the outlets.
    b. Proportionally balance the common ducts. Some systems have manual balancing dampers in both the hot and cold duct before they combine. Other systems have only one manual balancing damper located in the common duct. Fig. 5.3
    c. Adjust manual zone dampers or automatic static pressure dampers for correct airflow.

5. Change fan speed as needed.

6. Take final readings and check the system in the heating mode.
    a. Proportionally balance the common ducts. Some systems have manual balancing dampers in both the hot and cold duct before they combine. Other systems have only one manual balancing damper located in the common duct. If separate dampers are provided adjust the hot duct damper

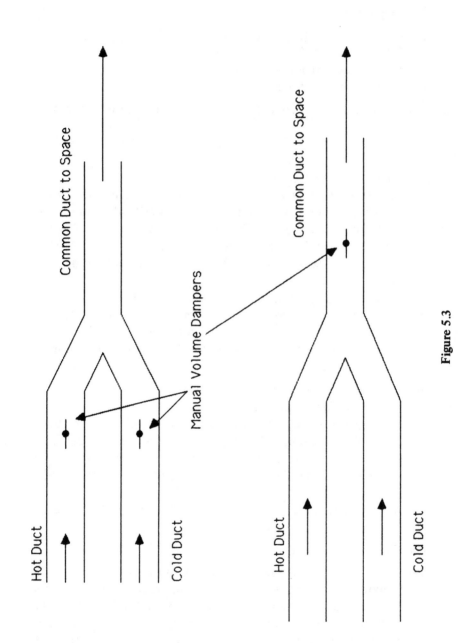

Figure 5.3

for correct cfm. If only one damper is provided *do not change from cooling setting.*

b. Adjust manual zone dampers or automatic static pressure dampers for correct airflow.

c. Read the outlets. The air volume should be close to cfm read when balanced in the cooling position. If it's not, check the system starting with the space thermostat and mixing dampers.

7. Check the system in the maximum outside air mode. If the motor overloads or the flow is excessive, adjust outside air dampers or fan speed as needed.

8. Complete report. Pressure dependent systems can't be controlled, nor will they maintain their balance, as well as pressure independent systems.

INDUCTION SYSTEMS

Induction systems are constant or variable volume, pressure independent systems. The induction units are supplied with medium to high pressure primary air through risers generally located around the perimeter of the building.

Balancing Procedure
For Constant Volume Induction Systems

1. Do the preliminary office work.
   a. Gather plans and specifications.
   b. Prepare report forms.

2. Do the preliminary field inspection.
   a. Inspect the job site.
   b. Inspect the distribution system.
   c. Inspect the equipment.

3. Make initial tests of all the equipment applicable to the system being balanced.
   a. Set all dampers at the full open position except the outside air which is set at minimum position.
   b. Check fan rotation.
   c. Take electrical measurements.
   d. Take speed measurements.

    e. Operate all fans at or near design speeds.

    f. Take static pressure measurements at the unit.

4. Balance and adjust the distribution system.

    a. Check the side of the induction terminal for a "cfm versus static pressure" chart. If not available, contact the terminal manufacturer for the data.

    b. Use a dry-type air pressure differential gage (Magnehelic) to read the primary air pressure at a typical nozzle in the first and last induction terminal on each riser. Determine the airflow being delivered but do not adjust these terminals.

    c. After all the risers have been read in the above manner, determine which risers have the highest pressures/airflow. Reduce the airflow to the risers with high pressures and continue the balance until all risers are close to being equal. Adjust the fan speed, if needed, so that the static pressure to the riser with the lowest pressure is adequate but not excessive.

    d. Starting with the induction terminals closest to the fan, adjust each terminal to within +/− 10% of design airflow. Use the terminal manufacturer's "cfm versus static pressure data chart/curve" or the following equations to convert static pressures to cfm.

$$\left(\frac{CFM_2}{CFM_1}\right)^2 = \frac{SP_2}{SP_1}$$

$$SP_2 = SP_1\left(\frac{CFM_2}{CFM_1}\right)^2$$

$$CFM_2 = CFM_1\sqrt{\frac{SP_2}{SP_1}}$$

    e. Two to three adjusting passes are normally needed to balance induction systems. A final pass without any adjustments should be made to record the results.

5. Change fan speed as needed.

6. Check the system in the maximum outside air mode. If the motor overloads or the flow is excessive, adjust outside air dampers or fan speed as needed.

7. Complete report.

Chapter 6
Balancing Variable Air Volume Systems

VARIABLE VOLUME (VAV) SYSTEMS

Most VAV systems use a single duct which supplies a constant air temperature, generally between 55 and 60 degrees, to the terminal box. There are also some dual duct systems which use two single duct supplies to the boxes.

The air volume through the terminal box and into the space is varied to maintain the space temperature. As the boxes throttle back to allow less air into the space, static pressure builds up in the supply duct. A static pressure sensor, generally installed about two-thirds to three-quarters of the way from the fan to the end of the duct system, senses the duct static pressure and sends a signal back to a device which controls supply air volume.

The location of the sensor is a compromise between energy efficiency and control of the system. For the greatest energy efficiency the sensor would be installed just before the terminal farthest (in terms of pressure loss) from the fan. The static pressure would be set only high enough to operate that box and its associated low pressure system. However, because the entire system is always changing, i.e., some boxes are closing while others are opening, the location of the farthest box is also constantly changing. Therefore, the sensor is located closer to the fan so that it represents the average static pressure in the system. However, the controller must be set at a higher static pressure to accommodate the additional pressure losses in the main ductwork downstream of the sensor. This gives control to the system but also increases the static pressure that the fan must produce, and some energy savings is lost.

After the system has been proportionally balanced and set up for diversity, you'll need to check the sensor for correct location and the controller for proper operation. This can be done by setting the system to maximum flow and static pressure and then observing the static pressure changes at the controller, the fan discharge and the sensor as some boxes are changed over from minimum to maximum. Next operate the system from maximum airflow to minimum airflow. Vary the sequence of changes to get a valid profile of how the system is performing.

The volume control device which receives the signal from the sensor may be (1) an electronic variable frequency drive which varies the speed of the motor and the fan speed, (2) a mechanical variable speed drive which changes the pitch of the motor sheave, varying the fan speed, (3) a variable pitch fan, (4) inlet or vortex dampers at the inlet of the fan, or (5) static pressure dampers at the fan discharge. The volume control devices are listed in order by energy efficiency with the first, second and third devices being comparable.

The exception to the above static pressure control scheme is the bypass type of system which uses relief dampers to bypass the air back to the fan inlet as the terminal boxes close. This type of VAV system has no energy savings as the primary fan is constant volume.

TYPES OF VAV SYSTEMS

Variable air volume systems can be pressure independent or pressure dependent. The pressure independent systems are: Single Duct, Dual Duct, Induction, Fan Powered, and System Powered. The pressure dependent systems are: Single Duct, Fan Powered and Bypass.

BALANCING PRESSURE INDEPENDENT SYSTEMS

Pressure independent systems have teminal boxes which work off the space thermostat signal as the master control. This signal operates a damper motor which in turn opens and closes the box's volume damper. A velocity controller is used as a submaster control to maintain the maximum and minimum air volume to the space. The maximum to minimum airflow will be maintained when the static pressure at the inlet of the box is in compliance with the box manufacturer's published data.

General Balancing Procedure

1. Do the preliminary office work.
   a. Gather plans and specifications.
   b. Prepare report forms.
2. Do the preliminary field inspection.
   a. Inspect the job site.
   b. Inspect the distribution system. Where applicable check that the duct has been leak tested.
   c. Inspect the equipment.
3. Make initial tests of all the equipment applicable to the system being balanced.
   a. Set all dampers at the full open position except the outside air. If applicable, set the fan inlet (vortex) dampers at minimum position.
   b. Check fan rotation.
   c. Take speed measurements.
   d. Start the fan and take electrical measurements with the volume control set for minimum air flow. Gradually increase flow to maximum. Take electrical measurements and observe the system for any adverse effects caused by the higher pressures.
   e. Set the system for full cooling.
   f. Operate all fans (supply, return and exhaust) at or near design speeds.
   g. Take static pressure measurements at the unit.
   h. Take measurements at several boxes at the end of the system to determine if the inlet static pressure is at or above the minimum required by the manufacturer for VAV box operation. Increase or decrease fan speed as needed.
   i. Take total air measurements. If the system has diversity, determine the diversity ratio. Place enough terminal boxes in a maximum flow condition (by setting the boxes' space thermostat to the lowest temperature) to equal the design cfm output of the fan. The remaining boxes will be set to minimum flow. The pattern of setting the boxes to maximum and minimum flow for testing and balancing should approximate normal operating conditions.

4. Balance and adjust the distribution system. Consider each terminal box and associated downstream low pressure ductwork as a separate, independent system.

   a. Verify the action of the thermostat (direct acting or reverse acting) and the volume damper position (normally closed or normally open). Verify the range of the damper motor as it responds to the velocity controller.

   b. Consult the box manufacturer's data for the required pressure drop range across the box. Then add the pressure losses needed for the low pressure distribution system downstream of the box (approximately 0.1 inch wg per 100 foot of duct (equivalent length) and 0.05 to 0.1 inch wg for the outlet). This is the total required inlet static pressure. Verify that the box is at maximum flow. Take the static pressure drop across the box and the inlet static pressure. They should be within the required range.

   c. Connect a differential pressure gage, such as a manometer, to the controller's pressure taps and read differential pressure. Use the manufacturer's published data (usually found on labels on the side of the box) to convert the pressure readings to cfm. The manufacturers use this method to preset their boxes at the factory.

   d. Field conditions may be such that the inlet duct configuration to the box may give an erroneous reading at the sensor. To verify the box's pressure tap reading, take a Pitot tube traverse of the low pressure duct off the boxes and a total of the outlets to confirm the box setting. This will provide the actual cfm delivered by the box and help to determine amount and location of any low pressure duct leakage.

   e. Proportionally balance the outlets.

   f. Set the box's controller for correct cfm using whichever of the following methods is appropriate: the box's pressure tap readings, the Pitot tube traverse, the total of the outlets, or a proportioned outlet.

   g. Set the box to minimum flow by setting the box's space thermostat above the space temperature. Check the total cfm and adjust the box's minimum setting following manufacturer's recommendations. Read and record the cfm of the individual outlets. They should remain in proportion.

However, it's normal for some outlets to be out of balance in the minimum setting. Do not rebalance. Leave the system balanced for maximum flow.

5. If the system pressure is low:
   a. Put enough boxes adjacent to the box being tested in the minimum flow position to bring the inlet static pressure at the test box to the required amount.
   b. Proportionally balance the outlets and set each box for maximum and minimum.
   c. After all the boxes have been set change fan speed as needed to get required system static pressure.
   d. Set the system for diversity if applicable.
   e. Take volume readings on the entire system.

6. Read and record the static pressure at the sensor.

7. Check the system in the maximum outside air mode. If the motor overloads or the flow is excessive, adjust outside air dampers or fan speed as needed.

8. Complete report.

Dual Duct Systems

Dual duct VAV systems may be dual or single fan. The single fan system has terminal boxes with two single duct variable inlets. The control scheme can be a deadband system using velocity controllers for each inlet in sequence with one another so the box supplies a varying amount of cooled or heated (but not mixed) air to the space. Another control variation has a velocity controller with sensor in either the hot or cold duct controlling airflow.

The dual fan system will have some boxes with both a hot and cold duct. These will generally supply exterior zones. Interior zones will have single inlet boxes. Dual fan systems may not have a heating coil. The heating fan is connected to, and supplies, return air only. The cooling fan supplies outside air and chilled air only. The control sequence uses a deadband scheme.

Balancing is similar to dual duct constant volume except where applicable, both hot and cold duct volume controllers must be set. Consult the box manufacturer for specific operating sequences.

Low pressure in the system can cause mixing in the two ducts resulting in excessive energy usage and poor space temperature control.

Induction Systems

Consult the box manufacturer for specific operating sequences. Aside from the general procedures, balancing induction systems will require setting the static pressure in the primary air duct to overcome the resistance of the box and the secondary ductwork to induce ceiling return air.

Fan Powered Systems

Fan powered terminal boxes use a modulating primary air damper and barometric return air damper in conjunction with a secondary fan to overcome the resistance of the low pressure ductwork and provide constant airflow to the space. Consult the box manufacturer for specific operating sequences. Balance the output of the secondary fan at design airflow when in the full return mode. In some fan powered boxes, the return air inlet backdraft damper may be a source of air leakage when the box is in the maximum cooling mode.

System Powered

System powered terminals are balanced the same as the standard VAV system. Again, consult the box manufacturer for specific operating sequences. One problem with system powered boxes that operate normally open is that sometimes they have difficulty building up enough air to operate the box's VAV controls. Upon startup, when the boxes are full open there may not be adequate static pressure in the system to control the boxes. This results in the outlets delivering excessive amounts of air until the static builds up and the boxes start controlling.

BALANCING
PRESSURE DEPENDENT SYSTEMS

Pressure dependent terminal boxes do not have an automatic volume controller to regulate airflow as the inlet static pressure changes as do the pressure independent boxes. What they do have is an automatic inlet volume damper controlled by the space thermostat.

These volume dampers may or may not have a minimum position limiter. The airflow delivered by the box is solely dependent on the inlet static pressure and therefore, changes as the inlet static changes. This type of system will also usually have a manual balancing damper at the inlet of the box and balancing dampers in the branch lines. The problem with pressure dependent systems, whether the system is being balanced or is in normal operation, is that every change in damper setting at one box affects adjacent boxes.

General Procedure — Non-Diversity Systems

1. Do the preliminary office work.
   a. Gather plans and specifications.
   b. Prepare report forms.
2. Do the preliminary field inspection.
   a. Inspect the job site.
   b. Inspect the distribution system. Where applicable check that the duct has been leak tested.
   c. Inspect the equipment.
3. Make initial tests of all the equipment applicable to the system being balanced.
   a. Check rotation.
   b. Take electrical measurements.
   c. Take speed measurements.
   d. Set all dampers at the full open position except the outside air which is set at minimum position.
   e. Set the system for full cooling.
   f. Operate all fans (supply, return and exhaust) at or near design speeds.
   g. Take static pressure measurements at the unit.
4. Balance and adjust the distribution system. Non-diversity systems are balanced similar to constant volume systems. Start with the box that has the lowest percent or design flow.
   a. Take a Pitot tube traverse of the low pressure duct off the box and the total of the outlets to determine duct leakage.
   b. Proportionally balance the outlets.
   c. Proportionally balance the terminals using the inlet volume damper.
   d. Proportionally balance the branches.

5. Change fan speed as needed.

6. Take final readings and check the system in the heating/minimum mode.

7. Check the system in the maximum outside air mode. If the motor overloads or the flow is excessive, adjust outside air dampers or fan speed as needed.

8. Complete report.

General Procedure — Diversity Systems

1. Do the preliminary office work.
    a. Gather plans and specifications.
    b. Prepare report forms.

2. Do the preliminary field inspection.
    a. Inspect the job site.
    b. Inspect the distribution system. Where applicable check that the duct has been leak tested.
    c. Inspect the equipment.

3. Make initial tests of all the equipment applicable to the system being balanced.
    a. Check rotation.
    b. Take electrical measurements.
    c. Take speed measurements.
    d. Set all dampers at the full open position except the outside air which is set at minimum position.
    e. Set the system for full cooling.
    f. Operate all fans (supply, return and exhaust) at or near design speeds.
    g. Take static pressure measurements at the unit. Place enough terminal boxes in a maximum flow condition (by setting the boxes's space thermostat to the lowest temperature) to equal the design cfm output of the fan. The remaining boxes will be set to minimum flow. The pattern of setting the boxes for balancing should approximate normal operating conditions.

4. Change fan speed as needed.

5. Balance and adjust the distribution system. Diversity systems are balanced opposite of non-diversity systems. Start with the box

with the highest percent of design airflow.

a. Take Pitot tube traverses of the low pressure duct off the box and a total of the outlets to determine duct leakage.

b. Proportionally balance the outlets.

c. Cut the box to 100% of design airflow using the manual volume damper at the inlet.

d. Continue adjusting boxes that are over 100% of design cfm until all the boxes which have been set for diversity are at 100% of design. If required, adjust the minimum setting for each of these boxes.

e. Next, start with the boxes closest to 100% of design cfm. Set adjacent boxes to minimum until enough air is available to bring the test box to within +/− 10% design flow. *Do not change any of the inlet balancing dampers already set.* Proportionally balance the outlets of the test box and set the box for minimum flow, if applicable. Even if some boxes can't be brought to design cfm, continue to proportionally balance the outlets and record the readings.

6. Read and record the static pressure at the sensor.

7. Take final readings.

8. Check the system in the maximum outside air mode. If the motor overloads or the flow is excessive, adjust outside air dampers or fan speed as needed.

9. Complete report.

COMBINATION SYSTEMS

When combination systems are encountered balancing procedures have to be modified to fit the pecularities of each system. Contact the box manufacturer for help. As a rule, it's usually best to balance the pressure independent terminals first, since they shouldn't be affected by changes in inlet static pressure. If the system doesn't have adequate static pressure for balancing the pressure independent boxes, set adjacent pressure independent boxes to minimum flow and close the inlet balancing dampers of the adjacent pressure dependent boxes. Set the test box for maximum flow and proportionally balance the outlets. After setting all the pressure independent boxes, set the pressure dependent terminals and balance their outlets.

Variable Primary/Variable Secondary (Master/Slave Units)

In variable primary/variable secondary systems the master unit has a pressure independent volume controller. The slave unit receives a signal from the master unit which positions the volume damper in the slave unit to the same position as the volume damper in the master unit. The master unit is balanced in the same manner as any pressure independent terminal. The slave unit is a pressure dependent unit and will deliver airflow in relation to its inlet static pressure. If the inlet static pressure at the slave is the same as at the master, the airflow will also be the same. However, if the inlet static at the slave is different from the master, the airflow will be different. Therefore, the master and all slave units should be on the same duct system. There's no adjustments or balancing of the slave units.

VAV TROUBLESHOOTING GUIDELINES

Diffuser Dumps Cold Air
Airflow too low (velocity too slow).
Check to determine if box is reducing too far.
Evaluate box minimum setting.
Diffuser is too large; check installation.

Conditioned Space Is Too Cold
Supply air temperature is too cold.
Too much supply air.
Diffuser pattern or throw is incorrect causing drafts.
Temperature sensor is located incorrectly or needs calibration.

Conditioned Space Is Too Warm.
Supply air temperature setting is too warm.
Not enough supply air.
Refrigeration system not operating properly.
Fan-coil evaporator is iced over because of low airflow.
Temperature sensor is located incorrectly or needs calibration.
Low pressure duct leaking.
Low pressure duct not insulated.

Cold air from diffuser isn't mixing properly with room air; increase
   volume or velocity, change or retrofit diffuser.

Noise.
Too much air in low pressure duct; check box maximum setting.
Static pressure in the system is too high.
Diffuser is too small.
Diffuser is dampered at face; always damper at takeoff.
Pattern controllers loose; tighten or remove.

Not Enough Air
Box not operating properly; check minimum setting; reset as neces-
   sary.
Not enough static pressure at box inlet for proper operation.
Damper in VAV box is closed; may be loose on shaft or frozen.
Low pressure damper closed.
Restrictions in low pressure duct.
Remove pattern controllers in diffusers.
Low pressure duct is leaking, disconnected or twisted.
Install VAV diffusers where applicable; these diffusers close down at
   the face producing higher air velocities as they reduce in size.
Install fan powered boxes.

Box Not Operating Properly
Not enough static pressure at the inlet.
Too much static pressure at the inlet.
Static pressure sensor is defective, clogged, or located incorrectly.
Static pressure setting on controller is incorrect.
Static pressure controller needs calibrating.
Fan speed not correct.
Inlet duct leaking or disconnected.
Main ductwork improperly designed.
Not enough straight duct on the inlet of the box.
Diversity is incorrect.
Box is wrong size or wrong nameplate. Check installation. Leak test
   if necessary.
Damper loose on shaft.
Linkage from actuator to damper is incorrect or binding.
Actuator is defective.

Controls are defective, need calibration or are set incorrectly.

Volume controller not set properly for normally open or normally closed operation.

Damper linked incorrectly; NO for NC operation or vice versa.

Pneumatic tubing to controller is piped incorrect, leaking, or pinched.

Restrictor in pneumatic tubing is missing, broken, placed incorrectly, wrong size, or clogged.

Oil or water in pneumatic lines.

No power to controls.

Wired incorrectly.

PC board defective.

Fan Not Operating Properly

Inlet vanes on centrifugal fans not operating properly.

Pitch on vaneaxial fans not adjusted correctly.

Fan rotating backwards.

Return air fan not tracking with supply fan. Check static pressure sensors, airflow measuring stations and move, clean or calibrate. Consider replacing return fan with a relief fan.

Negative Pressure in The Building
(Office buildings should be maintained at +0.03 to +0.05 inches of water.)

1. Caused by stack effect or improper return air control.

   Seal building properly.

   Balance return system, and install manual balancing dampers as needed to control OA, RA and EA at the unit.

   Get return fan to track with supply fan; consider replacing return fan with relief fan.

   Check that static pressure sensors are properly located and working.

   Install pressure controlled return air dampers in return air shafts from ceiling plenums.

2. Caused by the supply fan reducing air volume and a reduction of outside air for (1) the constant volume exhaust fans and (2) exfiltration.

   Increase minimum outside air by:

   a. Opening manual volume damper in outside air duct and/or increasing outside air duct size.

   b. Control OA from supply fan. As fan slows, outside air damper opens.

c. Control OA damper from flow monitor in OA duct to maintain a constant minimum OA volume.

Inadequate Amount of
Outside Air for Proper Ventilation

Caused by the supply fan reducing air volume and a reduction of outside air for code or building requirements.

Increase minimum outside air by:

a. Opening manual volume damper in outside air duct and/or increasing outside air duct size.

b. Control OA from supply fan. As fan slows, outside air damper opens.

c. Control OA damper from flow monitor in OA duct to maintain a constant minimum OA volume.

Chapter 7
Setting Outside Air
And Changing Fan Speeds

OUTSIDE AIR CONSIDERATIONS

Generally, because of poor locations, measurement of outside air quantities is extremely difficult. To make setting outside air quantities easier some systems have a constant volume regulator or a flow measuring station with automatic and balancing dampers in the outside air intake. If the system is variable air volume, a fan may be installed in the outside air intake or in the exhaust duct to bring in the proper amount of outside air to maintain the building under a slightly positive (commercial buildings are normally set between 0.03" to 0.05") pressure and to meet exhaust air and fresh air requirements when the fan is at minimum flow.

SETTING OUTSIDE AIR

After the terminals are balanced to the specifications, go back to the unit and reset the minimum outside air dampers. Outside air may be determined either by (1) traversing the outside air duct, (2) subtracting the measured return air cfm from the measured supply air cfm, (3) reading the outside, return, mixed or supply air temperatures.

The first two methods are preferred because temperature readings generally are both difficult to take accurately and time consuming. However, if the other methods aren't satisfactory, take temperature readings, average them and then use the following equations to set the outside air dampers:

Equation 7.1 $\text{MAT} = (\%\text{OA} \times \text{OAT}) + (\%\text{RA} \times \text{RAT})$

Equation 7.2 $\%\text{OA} = \dfrac{(\text{MAT} - \text{RAT}) \times 100}{(\text{RAT} - \text{OAT})}$

Where:

MAT   = Mixed Air Temperature
%OA   = Percent of Outside Air cfm
OAT   = Outside Air Temperature
%RA   = Percent of Return Air cfm
RAT   = Return Air Temperature

Problem 7.1: A fan supplies 15,000 cfm. The required outside air is 2250 cfm. The outside air temperature is 98 degrees. The return air is 80 degrees. Determine the mixed air temperature which corresponds to the correct quantity of outside air and set the dampers.

Answer: 82.7

Solution:

1. Solve for quantity of outside air. $2250/15000 = 0.15$ or 15%

2. Solve for MAT. $MAT = (\%OA \times OAT) +$
                                        $(\%RA \times RAT)$
                                        $MAT = (15\% \times 98) + (85\% \times 80)$
                                        $MAT = 14.7 + 68$
                                        $MAT = 82.7$

3. Adjust the outside dampers until an average temperature of 82.7 degrees is measured in the mixed air plenum. The outside air dampers are now set for 15%, or 2250 cfm of outside air.

One way to determine average temperature is to use a digital single or multiple sensor thermometer and take a temperature traverse in the plenum at the center of each filter. Take temperatures and set the dampers when the temperature difference between the outside air and return air is greatest. During the temperature traverse you may find some air stratification which will result in uneven heat transfer across the coil and can lead to coil freezeup or the unit shutting down when the freezestat trips. To correct air stratification, install baffles or some other type of air mixing device.

If a reliable mixed air temperature can't be obtained, the fan discharge air temperature can be used with the outside air temperature and the return air temperature to determine the mixed air quantities. The fan will add about ½ to ¾ degree Fahrenheit per inch of measured total static pressure to the air because of compression and motor heat. Therefore, an amount equal to this temperature rise across the fan must be subtracted from the discharge air temperature.

Equation 7.3 Determining fan discharge air temperature.

$$FDAT = (\%OA \times OAT) + (\%RA \times RAT) + 0.5(TSP)$$

Equation 7.4 Determining percent of outside air.

$$\%OA = \frac{RAT - [FDAT - 0.5(TSP)]}{RAT - OAT} \times 100$$

Where:
FDAT = Fan Discharge Air Temperature
%OA   = Percent of Outside Air cfm
OAT   = Outside Air Temperature
%RA   = Percent of Return Air cfm
RAT   = Return Air Temperature
TSP   = Total Static Pressure rise across the fan, in. wg
0.5   = ½ degree per inch of total static pressure

CHECKING ECONOMIZER OPERATION

A full air-side economizer automatically varies the position of the outside air and return air dampers (and sometimes exhaust air dampers) to maintain the proper mixed air temperature for the most economical operation of the HVAC system. It works like this: when the system is calling for cooling, outside air that's below the design room temperature is brought into the system and the mechanical cooling is cycled down or off, resulting in "free" cooling.

To check the operation of the economizer cycle, move the outside air and return air dampers from minimum outside air (the return air dampers should be full open) to maximum outside air (the return air dampers should be full closed) and back while observing the static pressures in the mixed air plenum. Any change in static pressure indicates a change in airflow quantity. Set the outside air controller to bring in maximum outside air, and then operate the mixed air controller so the dampers can be moved and stopped at 10% increments. Take a mixed air static pressure for each point of damper operation. This will give a static pressure profile of the economizer damper operation and confirm that the automatic dampers operate simultaneously to close the return air as the outside air is opening. Lagging, or leading damper operation, is a common cause of reduced airflow resulting from the chocking affect of one damper closing before the other opens. Correct any problems.

Generally speaking, the damper used for maximum outside air shouldn't be used for minimum outside air. A separate, properly sized damper should be used for minimum outside air. Also, the minimum outside air damper should be backed up by a constant volume regulator or manual damper so the proper amount of outside air can be set.

CALCULATING FAN SPEED CHANGES

If you find that a speed change is needed to bring the airflow to within $+/-10\%$ of design, use the fan laws to calculate fan speed (fan law No. 1) and horsepower (fan law No. 3) requirements. After determining the fan speed needed, use the following drive equation to determine the size of the sheaves needed to get the correct fan speed and airflow.

Equation 7.5: Drive equation RPMm x Dm = RPMf x Df

7.5.1 RPMm = RPMf x Df/Dm

7.5.2 Dm = RPMf x Df/RPMm

7.5.3 RPMf = RPMm x Dm/Df

7.5.4 Df = RPMm x Dm/RPMf

Where:

RPMm = speed of the motor shaft

Dm   = pitch diameter of the motor sheave

RPMf = speed of the fan shaft

Df   = pitch diameter of the fan sheave

From the drive equation:

1. Increasing the size of a fixed pitch motor sheave, or adjusting the belts to ride higher in a variable pitch motor sheave, will result in an increase in fan speed.

2. Decreasing the size of a fixed pitch motor sheave, or adjusting the belts to ride lower in a variable pitch motor sheave, will result in a decrease in fan speed.

3. Increasing the pitch diameter of the fan sheave decreases the fan speed.

4. Decreasing the pitch diameter of the fan sheave increases the fan speed.

Pitch Diameter

Notice that pitch diameter is used in the calculations. For field calculations use the outside diameter of a fixed sheave for pitch diameter. For variable pitch sheaves, when the belt is riding down in the groove an approximation of the pitch diameter will be used for calculation purposes.

Example Problems

Problem 7.2: A variable pitch motor sheave has a pitch diameter of 5.8″. The fan sheave has a pitch diameter of 10.4″. The motor speed is 1750 rpm. Find the fan speed.

Answer: 976 rpm

Solution: Use equation 7.5.3
     RPMf = RPMm x Dm/Df
     RPMf = 1750 x 5.8/10.4
     RPMf = 976

Problem 7.3: A fan is delivering 15,000 cfm. The fan speed is 600 rpm. The fan sheave is 15.5″Pd. It's been decided that the output of the fan should be reduced to 13,500 cfm. To reduce the fan output, the fan speed will be decreased to 540 rpm. The motor speed is 1725 rpm and has a variable pitch sheave. The belt is riding at a pitch diameter of 5.4″. Find the new pitch diameter of the motor sheave to reduce the fan speed to 540 rpm.

Answer: 4.85″

Solution: Use equation 7.5.2
     Dm   = RPMf x Df/RPMm
     Dm   = 540 x 15.5/1725
     Dm   = 4.85

Problem 7.4: Both the motor and the fan have fixed sheaves. The motor sheave is 5.0″ Pd and the fan sheave is 10.4″ Pd. The fan speed is measured at 840 rpm and the motor is rated at 1750 rpm.

The airflow of the fan is increased to 12,500 cfm from 11,500 cfm. It's decided to change the fan sheave to increase the fan cfm. The new fan speed will be 913 rpm. Find the size of the new fan sheave.

Answer: 9.58"

Solution: Use equation 7.5.4

Df   = RPMm x Dm/RPMf
Df   = 1750 x 5.0/913
Df   = 9.58

TIP SPEED

Fans are built to withstand the increased stress of centrifugal force caused by increases in fan speed to a certain limit designated by the fan class. If the fan is rotating too fast for its stress limits, the wheel could fly apart or the fan shaft could be bent. Therefore, before increasing the fan speed, check the fan performance table or consult the fan manufacturer for the maximum allowable fan speed to ensure that the new operating conditions won't require a higher class fan.

Equation 7.6 Tip Speed

$$TS = \frac{Pi \times D \times rpm}{12}$$

Where:

TS   = Tip Speed, in feet per minute
Pi   = 3.14
D   = fan wheel diameter in inches
rpm   = revolutions per minute of the fan

Problem 7.5: Find the tip speed of a fan operating at 840 rpm and having a wheel diameter of 38 inches.

Answer: 8352 fpm

Solution:

$$TS = \frac{Pi \times D \times rpm}{12}$$

$$TS = \frac{3.14 \times 38 \times 840}{12}$$

ADDITIONAL DRIVE INFORMATION

After calculating new sheave size, the following information will be needed.

1. Motor or fan shaft diameter. Motor shaft diameters are in increments of 1/8" and fan shaft diameters are in increments of 1/16".

2. Distance between the centers of the shafts.

3. Bushing size. Sheaves may have a fixed bore in which case they fit the exact size of the shaft, or they may have a larger bore to accept bushings of various bore diameters to fit different shaft sizes.

4. Number of belt grooves.

5. Belt size.

6. Allowance for motor movement on the motor slide rail to allow for adjustment of belt tension.

CALCULATING BELT SIZE

After the size of the sheave is calculated, it'll be necessary to calculate the new belt length needed for the drives to determine if the belt(s) will also need changing. If it's necessary to change any of the belts on a multiple groove sheave, buy a matched set of belts. Otherwise, because belt lengths and tension strengths vary, some belts could end up being too tight and others too loose, resulting in excessive wear.

Equation 7.7 Equation for finding belt length.

$$L = 2C + 1.57 \,(D + d) + \frac{(D - d)^2}{4C}$$

Where:

  L   = belt pitch length
  C   = Center to center distance of the shafts

D   = pitch diameter of the large sheave
d   = pitch diameter of the small sheave
1.57 = constant (Pi/2)

Problem 7.6: A motor sheave has a pitch diameter of 11". The fan sheave pitch diameter is 18". The distance between the shaft centers is 25". Find the belt length needed.

Answer: This drive set will need one A95 belt which has an outside diameter of 97.2" and a pitch diameter of 96.3".

Solution:

$$L = 2C + 1.57(D + d) + \frac{(D - d)^2}{4C}$$

$$L = 2(25) + 1.57(18 + 11) + \frac{(18 - 11)^2}{4 \times 25}$$

$$L = 96.02$$

CHANGING DRIVE COMPONENTS

Removing Belts

Belts shouldn't be too tight or too loose. Slack belts will squeal on start-up, and they'll wear out quicker and deliver less power. Belts with excessive tension will also make belts fail earlier and can cause excessive wear on shaft bearings and possible overload of the motor and drive. The correct operating tension is the lowest tension at which the belts will perform without slipping under peak load conditions. To install belts:

1. Loosen and slide the motor forward.

2. Put the belts on the sheaves and move the motor back to adjust the belts for proper tension.

3. Secure the motor.

Removing, Mounting or Adjusting Sheaves

For proper removing or mounting of sheaves or adjustment of variable pitch sheaves, consult the manufacturer's published data. *Caution:* Before trying to remove or adjust the pitch diameter of a sheave be sure to loosen all locking screws.

Sheave Alignment

To prevent unnecessary belt wear or the possibility of a belt jumping off the sheaves, the motor and fan shafts should be parallel and the sheaves in alignment. To align the motor and fan sheaves:

1. Place a straightedge from the fan sheave to the motor sheave. The straightedge is on the outside flange of the sheaves.

2. Move the motor or the sheaves for equal distance from the straightedge to the center of both fan and motor sheaves.

Recheck

Drive components, belt tension and drive alignment should be rechecked after the first day's operation and again a few days later.

FINAL DATA

After the outside air has been set and any speed changes to the fan have been made, retest rpm, amperage and static pressure and make a final pass of the terminals, recording these readings on the data sheets. Compare final outlet readings with traverses; if there are no differences, mark the dampers so they can be reset if tampering or accidental changes occur. Verify that all automatic control dampers are sequencing properly.

Walk the system to determine if there are any drafts. If some are found they usually can be corrected by adjusting the outlet's pattern control device.

Review the report and ensure that nothing has been omitted and that all notes on problems, deficiencies, and other abnormal conditions are fully explained.

Chapter 8
Balancing Return Air And Toilet Exhaust Systems

RETURN AIR SYSTEMS

Frequently, return air systems aren't balanced. This is particularly true when there's only a supply fan and no return air fan. The philosophy is that if the supply air system, including the outside air is working properly the return air will take care of itself. In most cases I'd agree. However, the duct still should be inspected for air leakage, poor design or poor installation which would reduce or restrict proper airflow from the space. Also, systems with return fans, at the very least, will need to have the fans tested even if the return duct and inlets aren't proportionally balanced.

Return Air Balancing Procedure—
Systems with Return Air Fan

1. Gather plans and specifications.
   a. Prepare report forms.

2. Inspect the return air duct system.

3. Inspect the building.
   a. All windows, doors, etc., should be in normal position.

4. Set all return air dampers to full open position.
   a. The outside air damper is set at minimum position.

5. Test the return air fan.
   a. Confirm that the return air fan has correct rotation.
   b. Take electrical measurements.
   c. Take speed measurements.

6. Operate all fans associated with the return air system. The supply system has been balanced.

7. Take static pressure measurements at the return fan.

8. Take total air measurements in the return main.

9. Balance and adjust the distribution system.
   a. Read the return air inlets.
   b. Total the inlets and compare the total air at the fan. Reconcile any differences.
   c. Proportionally balance the inlets starting with the inlet with the lowest percent of design flow.
   d. Proportionally balance the branches.

10. Change fan speed as needed.

11. Take final readings.

12. Check the system in the maximum outside mode. If the motor overloads or the flow is excessive, adjust the exhaust air dampers (economizer systems) or the return fan speed as needed.

13. Complete reports.

Return Air Balanacing Procedure —
Systems without Return Air Fan

1. Gather plans and specifications.
   a. Prepare report forms.

2. Inspect the return air duct system.

3. All windows, doors, etc., should be in normal position.

4. Set all return air dampers to full open position. The outside air damper is set at minimum position.

5. Set the supply fan for normal operation. The supply air system has been balanced.

6. Take total air measurements in the return main.

7. Balance and adjust the distribution system.
   a. Read the return inlets. Total the inlets and compare with the total air in the return main. Reconcile any differences.
   b. Proportionally balance the inlets starting with the inlet with the lowest percent of design flow.
   c. Proportionally balance the branches.

8. Take final readings.

9. Complete reports.

Balancing Procedure —
Ceiling Return Systems with Return Air Fan

1. Gather plans and specifications. Prepare report forms.

2. Inspect the return air system.

3. All windows, doors, etc., should be in normal position.

4. Set all return air dampers to full open position. The outside air damper is set at minimum position.

5. Test the return air fan.
   a. Confirm that the return air fan has correct rotation.
   b. Take electrical measurements.
   c. Take speed measurements.

6. Operate all fans associated with the return air system. The supply system has been balanced.

7. Take static pressure measurements at the return fan.

8. Take total air measurements in the main.

9. Balance and adjust the distribution system.
   a. Take Pitot tube traverses of the return air duct in the ceiling return or take a rotating anemometer reading at the duct openings and set volume dampers.

10. Change fan speed as needed.

11. Take final readings.

12. Check the system in the maximum outside mode. If the motor overloads or the flow is excessive, adjust the exhaust air dampers (economizer systems) or the return fan speed as needed.

13. Complete reports.

Balancing Procedure —
Ceiling Return Systems without Return Air Fan

1. Gather plans and specifications. Prepare report forms.

2. Inspect the return air system.

3. All windows, doors, etc., should be in normal position.

4. Set all return air dampers at full open position. The outside air damper is set at minimum position.

5. Set the supply fan for normal operation. The supply air system has been balanced.

6. Take total air measurements in the return main.

7. Balance and adjust the distribution system.

8. Take final readings.

9. Complete reports.

TOILET EXHAUST SYSTEMS

There are many installation and design problems associated with toilet exhaust systems. Some problems are:

1. Volume dampers not installed.

2. Backdraft dampers not installed or won't open or close properly.

3. Excessive length of duct runs.

4. Excessive pressure drops at the fan inlet resulting from poorly designed ductwork.

5. Excessive duct leakage.

6. Using the building structure, such as masonry shafts, instead of ductwork as the conduit for the exhaust air.

Balancing Procedure

1. Gather plans and specifications.
   a. Prepare report forms.

2. Inspect the exhaust air duct system.

3. All windows, doors, etc., should be in normal position.

4. Test the exhaust air fan.
   a. Confirm that the fan has correct rotation.
   b. Take electrical measurements.
   c. Take speed measurements.

5. Take total air measurements in the exhaust main.

6. Balance and adjust the distribution system.

    a.   Read the system. Total the inlets and compare the total air at the exhaust fan. Reconcile any differences.

    b. Proportionally balance the inlets starting with the exhaust inlet with the lowest percent of design flow.

    c.   Proportionally balance the branches.

7. Change fan speed as needed.

8. Take final readings.

9. Complete reports.

Chapter 9
Fans, Fan Performance Curves
And Duct System Curves

FAN OPERATION

As the fan wheel is rotated, centrifugal force causes the air in the fan to be thrown outward from the wheel which reduces the pressure at the inlet of the wheel, allowing more air to be forced in through the fan suction opening by atmospheric pressure. As the air leaves the wheel at a high velocity it's collected in the fan housing where the velocity is reduced and converted into pressure. This pressure developed by the centrifugal fan depends on the wheel diameter and speed of rotation and is entirely the result of the velocity imparted to the air by the wheel.

FAN DESIGNATIONS

Fans are designated by type, class, fan rotation, drive arrangement, motor location, air discharge direction and, for centrifugal fans, width of fan wheel (single wide or double wide). For balancing purposes it's important to understand and be able to recognize fan types and their performance characteristics, class and rotation.

FAN TYPES

HVAC fans are divided into three general categories: axial fans, centrifugal fans, and special design fans such as tubular centrifugal fans, centrifugal power roof ventilators and axial power roof ventilators.

AXIAL FANS

The three general classifications of axial fans are: propeller, tubeaxial and vaneaxial.

Propeller Fans

Duty: Limited to low pressure. Propeller fans are best used for delivering large volumes of air at low pressures. A typical application would be general air circulation or exhaust ventilation without any attached ductwork.

Housing: Simple ring enclosure.

Fan Wheel Design: Wheels usually have two or more single thickness blades.

Discharge: The air is discharged in a circular or spiral pattern.

Efficiency: Generally low. Maximum efficiency is reached near free delivery.

Horsepower Characteristics: Lowest at maximum air delivery. Highest at minimum air delivery. Horsepower increases as static pressure increases.

Flow Characteristics: Airflow within the wheel is parallel to the shaft.

Static Pressure Range: Generally ¾ in. wg or less.

Tubeaxial Fans

Tubeaxial fans are heavy duty propeller fans.

Duty: Medium pressures. Tubeaxial fans are used in ducted HVAC applications where noise and air distribution on the downstream side aren't critical, such as fume exhaust systems, paint spray booths and drying ovens.

Housing: The wheel is enclosed in a cylindrical tube to increase efficiency and pressure capabilities.

Fan Wheel Design: The wheel is similar to the propeller type except it usually has more blades, 4 to 8, and they are heavier design, either airfoil or single thickness cross section.

Discharge: The air is discharged in a circular or spiral pattern which produces higher system losses than straight air flow.

Efficiency: Medium. Maximum efficiency is reached near free delivery. The tubeaxial fan has a significant improvement in efficiency and static pressure capability over propeller fans.

Horsepower Characteristics: Lowest at maximum air delivery. Highest at minimum air delivery. Horsepower increases as static pressure increases. Ensure that automatic dampers operate freely and take care not to overload the motor when closing volume dampers.

Flow Characteristics: Airflow within the wheel is parallel to the shaft. Medium to high air volumes.

Static Pressure Range: Typically to 3 in. wg.

Performance Curve Characteristics: The performance curve has a dip to the left of peak pressure and pressures in this area should be avoided. This is an undesirable characteristic because under certain conditions the fan may "aerodynamically" stall.

Vaneaxial Fans
Vaneaxial fans are tubeaxial fans with straightening vanes.

Duty: Vaneaxial fans are used in applications where good downstream air distribution is needed and space is a consideration while noise isn't.

Housing: Cylindrical tube with straightening vanes.

Fan Wheel Design: The vaneaxial fan wheel has shorter blades and a larger hub than the tubeaxial. The most efficient blades are airfoil design. Blades may be fixed or adjustable pitch. Adjust the pitch of the blades to vary the fan static pressure and cfm. Consult the manufacturer for adjustment instructions.

Discharge: The straigntening vanes straighten out the spiral motion of the air and improve pressure characteristics and efficiency of the fan.

Efficiency: High. Maximum efficiency is reached near free delivery.

Horsepower Characteristics: Lowest at maximum air delivery. Highest at minimum air delivery. Horsepower increases as static pressure increases. Ensure that automatic dampers operate freely and take care not to overload the motor when closing volume dampers.

Flow Characteristics: Airflow within the wheel is parallel to the shaft. Medium to high air volumes.

Static Pressure Range: Medium to high.

Performance Curve Characteristics: The performance curve includes a dip to the left of peak pressure which should be avoided. This is an undesirable characteristic because under certain conditions the fan may "aerodynamically" stall.

CENTRIFUGAL FANS

The four general classifications of centrifugal fans are: forward curved, backward curved or inclined, airfoil, and radial.

Forward Curved Centrifugal Fan

The forward curved fan is sometimes called a squirrel cage fan.

Duty: Generally used in low to medium pressure applications such as residences and in package units because it's quieter than the higher speed, higher pressure fans.

Housing: Light weight construction.

Fan Wheel Design: The fan wheel has 24 to 64 shallow blades with the blades curving toward the direction of rotation. The wheel is usually 24" in diameter or smaller. There may also be multiple wheels on a common shaft.

Discharge: Top Horizontal (TH), Bottom Horizontal (BH), Up Blast (UB), Down Blast (DB), Top Angular Down (TAD), Bottom Angular Down (BAD), Top Angular Up (TAU), and Bottom Angular Up (BAU). Fan discharge is always viewed from the drive side with the fan sitting on the ground.

Efficiency: Highest efficiency occurs to the right of peak pressure when the fan is delivering 40 to 50% of wide open volume.

Horsepower Characteristics: Minimum horsepower at no delivery. Horsepower curve increases continuously as the air quantity increases and the static pressure decreases. The forward curved fan is an "overloading" fan. The motor could overload if static pressure is reduced below design. An example is a fan picked to operate at 45% of wide open volume at 90% static pressure. The motor is chosen to operate at a maximum of 4.0 brake horsepower and is now operating at 3.5 brake horsepower. If for instance, the fan access door is removed, the system resistance would be reduced and the fan would handle more air at less static pressure. If the static pressure dropped to 70%, the cfm would increase to 60% and the brake horsepower would go to about 5.0 and the motor would stop on overload.

Flow Characteristics: Airflow within the wheel is radial to the shaft. Low to medium volumes.

Static Pressure Range: Low to medium pressure.

Performance Curve Characteristics: There's a dip in the pressure curve left of the peak pressure point. This is an undesirable characteristic of the forward curved fan because under certain conditions the same static pressure may result in the fan operating at a lower cfm and then at a higher cfm causing fan pulsations. The pressure curve is less steep and the efficiency of forward curved fans is somewhat less than that of airfoil or backward curved or inclined fans.

Backward Curved and
Backward Inclined Centrifugal Fans

Duty: Commercial and industrial HVAC applications. Used in industrial applications where dust might cause erosion to airfoil blades.

Housing: Medium to heavy weight construction.

Fan Wheel Design: The backward bladed fan wheel has 10 to 16 blades with the blades curving away from the direction of rotation. The backward inclined bladed fan wheel has 10 to 16 blades with flat blades which lean away from the direction of rotation.

Discharge: Top Horizontal (TH), Bottom Horizontal (BH), Up Blast (UB), Down Blast (DB), Top Angular Down (TAD), Bottom Angular Down (BAD), Top Angular Up (TAU), and Bottom Angular Up (BAU). Fan discharge is always viewed from the drive side with the fan sitting on the ground.

Efficiency: More efficient than forward curved fans but less efficient than airfoil fans.

Horsepower Characteristics: Minimum horsepower at no delivery. The horsepower curve increases with an increase in air quantity but only to a point to the right of maximum efficiency and then gradually decreases. The backward curved and backward inclined fans are "non-overloading" fans. If a motor is selected to handle the maximum brake horsepower as indicated by the performance curve, it won't be overloaded in any condition of fan operation.

Flow Characteristics: Airflow within the wheel is radial to the shaft. Medium to high volumes.

Static Pressure Range: Medium pressure.

Performance Curve Characteristics: The backward curved or backward inclined fans don't have a dip in the pressure curve left of the peak pressure point as do the forward curved fans and therefore, give a more stable and predictable operation.

Airfoil Centrifugal Fan

Duty: Commercial and industrial HVAC applications.

Housing: Medium to heavy weight construction.

Fan Wheel Design: The airfoil fan wheel has 10 to 16 aerodynamically shaped blades similar to an airplane wing which curve away from the direction of rotation.

Discharge: Top Horizontal (TH), Bottom Horizontal (BH), Up Blast (UB), Down Blast (BD), Top Angular Down (TAD), Bottom Angular Down (BAD), Top Angular Up (TAU), and Bottom Angular Up (BAU). Fan discharge is always viewed from the drive side with the fan sitting on the ground.

Efficiency: The best efficiency of all the centrifugal fans with highest efficiencies occurring at 50 to 65% of wide open volume.

Horsepower Characteristics: Minimum horsepower at no delivery. The horsepower curve increases with an increase in air quantity but only to a point to the right of maximum efficiency and then gradually decreases. Airfoil fans are "non-overloading" fans. If a motor is picked to handle the maximum brake horsepower shown on the performance curve, it won't be overloaded in any condition of fan operation.

Flow Characteristics: Airflow within the wheel is radial to the shaft. Medium to high volumes.

Static Pressure Range: Medium to high pressure.

Performance Curve Characteristics: The airfoil fan doesn't have a dip in the pressure curve left of the peak pressure point as does the forward curved fan and therefore, gives a more stable and predictable operation.

Radial Centrifugal Fan

Duty: Industrial applications such as waste collection and other types of material handling which call for high velocities and pressures.

Housing: Heavy construction.

Fan Wheel Design: The fan has 6 to 10 "paddle wheel" blades which are radial, at the leaving edge, to the center of the wheel. The fan wheel is generally heavily constructed with the blades being narrow but comparatively large in diameter. The wheel is sometimes coated with special materials.

Discharge: Top Horizontal (TH), Bottom Horizontal (BH), Up Blast (UB), Down Blast (DB), Top Angular Down (TAD), Bottom Angular Down (BAD), Top Angular Up (TAU), and Bottom Angular Up (BAU). Fan discharge is always viewed from the drive side with the fan sitting on the ground.

Efficiency: The least efficient of all centrifugal fans.

Horsepower Characteristics: Minimum horsepower at no delivery. Horsepower increases continuously as the air quantity increases and the static pressure decreases. The radial fan is an "overloading" fan.

Flow Characteristics: Airflow within the wheel is radial to the shaft. Medium to high volumes.

Static Pressure Range: High pressures.

Performance Curve Characteristics: The curve may have a dip to the left of peak pressure but not as great as with the forward curved fan and it's usually not enough of a break to cause difficulty.

SPECIAL DESIGN FANS

Tubular Centrifugal Fan

Tubular centrifugal fans are also known as inline centrifugal fans.

Duty: The straight-through configuration is a space saver as compared to a regular centrifugal fan and therefore, is primarily used in return systems where saving space is a consideration.

Housing: Cylindrical tube similar to a vaneaxial housing except the outer diameter of the wheel doesn't run as close to the housing.

Fan Wheel Design: The fan wheel can be either backward inclined or airfoil blade.

Efficiency: Lower than backward curved or backward inclined fans.

Horsepower Characteristics: Minimum horsepower at no delivery. The horsepower curve increases with an increase in air quantity but only to a point to the right of maximum efficiency and then gradually decreases. Tubular centrifugal fans are "non-overloading" fans. If a motor is selected to handle the maximum brake horsepower shown on the performance curve, it won't be overloaded in any condition of fan operation.

Discharge and Flow Characteristics: The air is discharged radially from the wheel and then changes directions by 90 degrees to flow through a guide vane section and then runs parallel to the fan shaft. Low to medium volumes.

Static Pressure Range: Low pressure.

Performance Curve Characteristics: The fan performance is similar to the backward bladed fans except for lower capacities, pressures and efficiencies.

FAN PERFORMANCE CURVES

A fan performance curve is a graphic representation of the performance of a fan from free delivery to no delivery. The air density, wheel size and fan speed are usually stated on the curve and are constant for the entire curve. The following fan characteristics may be plotted against cfm: static pressure (SP), static efficiency (SE), total pressure (TP), total efficiency (TE) which is also known as mechanical efficiency (ME), and brake horsepower (BHP). A fan performance curve is shown in figure 9.1.

Fan performance curves are developed from actual tests. The Air Movement and Control Association (AMCA), and other fan manufacturers have established procedures and standards for the testing and rating of fans. According to AMCA publications their testing procedure normally requires that the entire range of the fan's performance be tested from free delivery to no delivery. Both the discharge pressure and the inlet pressure is measured. These measurements are then mathematically translated to air volume and fan pressure. The test fan is usually driven by a dynamometer which

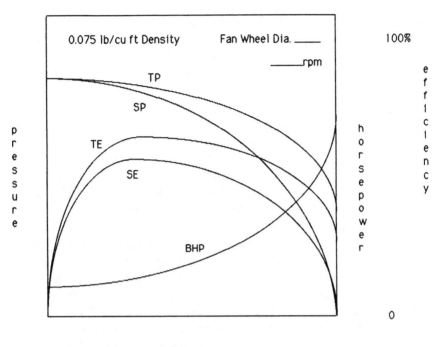

Figure 9.1
FAN PERFORMANCE CURVE

provides values of torque at each of the operating points while readings of fan speed are taken simultaneously. This allows calculation of horsepower input for each of the settings. Readings are also taken of dry and wet bulb temperature and barometric pressure so the air density can be calculated. These measured and calculated values are then plotted to develop the fan performance curve.

Although performance curves can be useful in troubleshooting fans, you should be aware that because of installation conditions which almost never duplicate the ideal conditions under which fans are tested, fan performance as determined by field tests is usually less than shown by manufacturers' tests.

FAN CLASSES

Most manufacturers offer fans in different classes. Classes are designated by number as Class I, II, III, or IV. The classifications are based on: fan construction and materials used, type of duty, static pressure developed, fan speed and outlet velocity. Each higher number represents higher speed and air performance capabilities of the fan.

FAN TABLES

To make fan selections and comparisons as simple as possible, fan manufacturers publish multi-rating or performance tables. These tables normally show the cfm, static pressure, rpm, outlet velocity, bhp, blade configuration, wheel configuration, fan wheel diameter, outlet area, tip speed equation, maximum brake horsepower equation, and pressure class limits for each class of fan.

The tables can also be used to help determine how the fan is operating under field conditions by measuring the fan speed and the fan static presssure and entering this information on the table. If the measured conditions are within the scope of the table the approximate cfm and brake horsepower can be determined.

Fans that have high rotating speeds and operate at high pressures are built to withstand the stresses of centrifugal force. However, if the fan is rotating too fast, the wheel could fly apart or the fan shaft could whip. Therefore, for safety reasons, the performance tables list the maximum rpm for each class of fan. Maximum allowable fan speed should be checked before increasing the fan rpm to ensure that the new operating conditons don't require a different class fan. The tip speed equation found on the performance table is only for that fan. To calculate the tip speed of any fan see Chapter 7, equation 7.6.

MULTIPLE FAN ARRANGEMENTS

Fans in Parallel

When fans operate in parallel the capacities and horsepowers are additive at equivalent pressures. For instance, two fans, each separately capable of handling 10,000 cfm at a static pressure of 2 in. wg, and 8 bhp, would if operating in parallel, move 20,000 cfm at 2 in. wg static pressure and 16 bhp.

If the fans in a parallel arrangement haven't been properly chosen or installed, one of the fans may handle less air than the other fan.

This unbalanced condition can result in the fan pulsating which will reduce the total performance and will cause noise and vibration and may damage the fan or the ductwork. Dampers near the inlets or the outlets may correct the unbalanced condition.

Fans in Series

When fans operate in series the pressures and horsepowers are additive at equivalent capacities. Using the example above, the two fans, each separately handling 10,000 cfm at 2 in. wg at 8 bhp would operate at 10,000 cfm at 4 in. wg and 16 bhp if placed in series.

DUCT SYSTEM CURVES

The system curve for any fan and duct system is a plot of the pressure needed to move the air through the system and overcome the total of all the pressure losses through the system components. For any point on the system curve, there's a given static pressure for a given cfm. For example, in figure 9.2, the static pressure of 1 in. wg corresponds to a cfm of 600 (point A). Any other point on the curve can be found by using fan law No. 2 (static pressure varies as the square of the cfm). For instance, point B on the curve is 1200 cfm and 4 in. wg SP. Point C is 1800 cfm and 9 in. wg SP.

Fixed System Curve

A fixed system is one in which there are no changes in the system resistance resulting from closing or opening of dampers, or changes in the condition of filters or coils, etc. For a fixed system, an increase or decrease in system resistance results only from an increase or decrease in cfm and this change in resistance will fall along the system curve shown in figure 9.2 If, however, dampers are operated toward their closed positions, or filters become dirty or cooling coils become wet from condensate, the system curve will no longer apply. Using field measured cfm and static pressure for the first point and then mathematically calculating the other points a new system curve may be plotted.

Operating Point

Each duct system has its own system curve and each fan at a

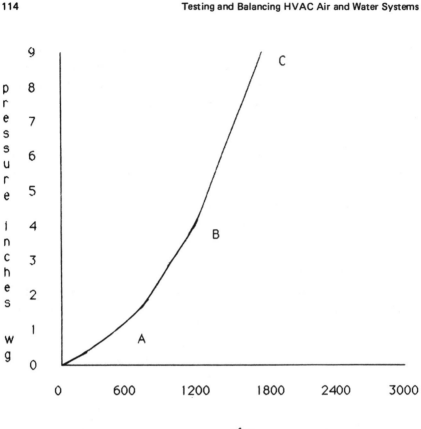

Figure 9.2

given speed has its own performance curve. The intersection of these two curves is the operating point for both the duct system and the fan. Just as every system operates only along its system curve; every fan operates only along its performance curve. For the duct system curve, the static pressure (system resistance) increases or decreases as the square of the cfm increases or decreases (fan law No. 2). However, for the fan, a decrease in static pressure (system resistance) will mean an increase in cfm and vice versa.

Using the fan laws, the fan performance curve, and the system curve, any change to the fan or the duct system can be calculated and graphically depicted. For example, in figure 9.3 the operating point for a fan operating on fan curve A in duct system I is point 1. To increase or decrease airflow a physical change must be made to either the duct system or the fan or both. If the change is to the fan, it will mean an increase or decrease in fan speed. After the speed change the fan will be operating on a new performance curve that runs parallel to the original curve. Since the duct system has remained unchanged, the system curve (I) also remains unchanged. In figure 9.3, an increase in fan speed results in a new fan curve (B). The fan and the system are now operating at a higher cfm and static pressure at point 2.

If, however, the increase or decrease in airflow is made by changing the duct system, by reducing or adding system resistance (for example, opening or closing a main damper), a new system curve is established while the fan performance curve (A) stays unchanged. In figure 9.3, an increase in system resistance results in a new system curve (II). The fan and system are now operating at a lower cfm and a higher static pressure at point 3.

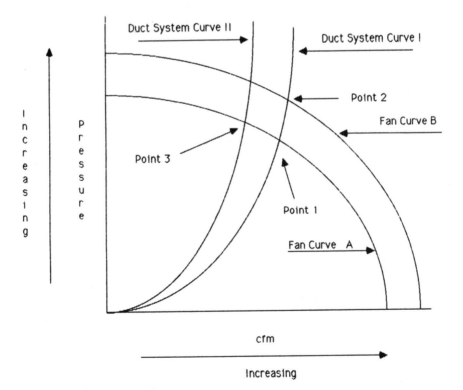

Figure 9.3

As already stated, fan measurements made in the field don't always fall into place on the performance curves. More often than not, there will be differences. This is due, in part, to many reasons such as the lack of good locations to take measurements and the way the duct is connected to the inlet and outlet of the fan. The duct configuration on the inlet and outlet of the fan has a marked effect on fan performance.

FAN PERFORMANCE

System Effect

System effect is the loss in performance of a fan resulting from adverse conditions at the fan inlet or outlet. These conditions change the aerodynamic characteristic of the fan so its full flow potential isn't realized. These conditions occur if the fan inlet or outlet connections aren't properly designed or installed to provide, as nearly as possible, uniform straight flow.

Since the velocity profile at the fan outlet isn't uniform, a length of straight duct is needed to establish a uniform velocity profile. Therefore, the outlet duct, including any transition duct, should extend out at least one duct diameter for each 1000 feet per minute of outlet velocity with a minimum length of 2½ duct diameters. For example, a fan with an outlet velocity of 1500 fpm would need a straight duct of 2.5 duct diameters, while a fan with an outlet velocity of 3000 fpm would need a straight length of duct of 3 duct diameters.

OUTLET CONNECTIONS

Outlet Duct and Transitions

AMCA Standard 210 specifies an outlet duct that's not greater than 107.5% nor less than 87.5% of the fan outlet area. It also calls for the slope of the outlet transition not being greater than 15% for converging transitions nor greater than 7% for diverging transitions.

Elbows

An elbow too close to the fan outlet can increase the air turbulence to such a degree that design airflow quantities downstream of the elbow can't be attained. Therefore, a length of straight duct (see system effect above) should be installed at the fan outlet to establish a uniform velocity profile approaching the elbow. If an elbow must be installed near the fan outlet, the ratio of elbow radius to duct diameter should be at least 1.5 to 1 (fig. 9.4).

Takeoffs

A length of straight duct (see system effect above) should be installed at the fan outlet to establish a uniform velocity profile approaching the takeoff.

Volume Dampers

Volume dampers installed at or near the fan outlet are manufactured with either opposed or parallel blades. Parallel-bladed dampers when partly closed divert the airflow to one side of the duct resulting in a non-uniform velocity profile beyond the damper. This may create airflow problems for takeoffs close to the damper. Opposed-bladed dampers are a non-diverting type of damper and are

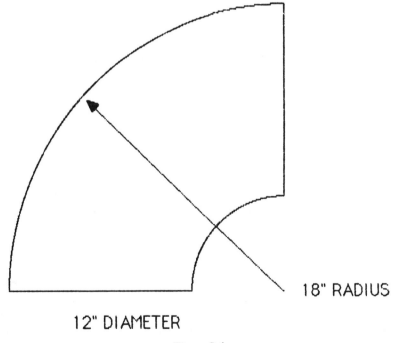

18" RADIUS

12" DIAMETER

Figure 9.4

therefore recommended for volume control at the fan outlet.

INLET CONDITIONS

Inlet Duct and Transitions

AMCA Standard 210 specifies an inlet duct that's not greater than 112.5% nor less than 92.5% of the fan inlet area. It also calls for the slope of the inlet transition not being greater than 15% for converging transitions nor greater than 7% for diverging transitions. To reduce losses caused by the reduction in flow area and then the following rapid expansion at the fan inlet, the inlet duct or fan inlet should have (1) a smooth, rounded entry, (2) a converging taper entry, or (3) a flat flange on the end of the duct.

Elbows

An elbow at the fan inlet won't allow the air to enter uniformly and will result in turbulent and uneven flow distribution at the fan wheel. Losses can be reduced if straight duct and turning vanes are added.

Inlet Spin

A major cause of reduced fan performance is an inlet duct condition that produces a pre- or counter-rotational spin to the entering air (fig. 9.5). These inlet conditions can be improved by installing turning vanes, splitters or airflow straighteners.

Fan Enclosures

Fan performance is reduced if the space between the fan inlet and the fan enclosure is too restrictive. Allow at least one-half the wheel diameter between the enclosure and the fan inlet. Inlets of multiple double wide centrifugal fans in parallel in a common enclosure should be at least one wheel diameter apart.

Belt Guards

Fan performance is reduced when a belt guard is in the plane of the fan inlet. Belt guards in the air stream should be made of open construction with as much free air passage as possible and shouldn't be more than one-third of the inlet area.

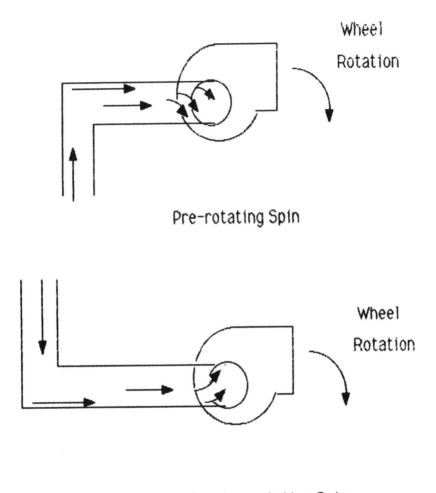

Counter-rotating Spin

Figure 9.5

Multiple Fans

If one of the fans in a parallel arrangement has a restricted inlet, it may handle less air than the other fan. This can result in the fan pulsating which will reduce the total performance and will cause noise and vibration and may damage the fan or the ductwork.

Chapter 10
Testing and Balancing
Fume Hood Systems

The laboratory work place must be safe for all the occupants of the lab facility. This includes laboratory personnel, maintenance people, support personnel, visitors, and anyone else within the fume hood or exhaust system environment. Although process function and occupancy comfort are extremely important they are secondary to safety. This chapter describes the basic types of laboratory fume hoods, and the procedures for testing hoods. Included are testing air volume and velocity, smoke testing, and troubleshooting laboratory fume hoods for reverse air flow and incorrect face velocities. Although the chapter emphasizes laboratory fume hoods, the testing procedures outlined, with few exceptions, are appropriate for kitchen and industrial fume hoods. A terminology section is at the end of this chapter.

LABORATORY HVAC SYSTEM

The laboratory space is supplied with filtered, conditioned air for temperature and humidity control by the supply air handling equipment. In the conditioned air, there is enough outside air to meet ventilation code requirements and maintain proper space pressurization. The supply air system and the exhaust air system may be either constant air volume, variable air volume, or a combination of the two. When the fume hood exhaust fan is energized, conditioned air from the laboratory space comes into the hood to contain and exhaust the contaminants generated inside the hood. The contaminants are ducted through the exhaust system to the outside where they are released into the atmosphere for dilution. To maintain laboratory temperature and pressurization, the exhausted air is generally replaced entirely by conditioned air from the supply system.

LABORATORY FUME HOODS

A laboratory fume hood is a ventilated, box-like structure that captures, contains, and exhausts contaminated fumes, vapors, and particulate matter generated inside the enclosure. Laboratory fume hoods are made of various materials such as epoxy-coated steel, epoxy resin, fiberglass, polypropylene, PVC and stainless steel.

The basic laboratory fume hood is mounted on a bench or table and has two side panels, a front, a back and a top panel. It also has an exhaust plenum with a baffle and an exhaust collar. The front of the hood is the face and is equipped with a movable, transparent sash. Sashes may be vertical-moving only, or a combination sash that has horizontally sliding panels set in a vertical-moving sash. For either type of sash, the vertical sash is placed in the highest position for easier setup or removal of laboratory apparatus in the hood. For normal hood use, other than setup or removal of apparatus, the vertical sash should be closed whenever someone is not using the hood. In any circumstance the sash should be opened only high enough to allow for proper use. With the combination sash, the vertical sash is closed completely and the horizontal sash is opened only wide enough for proper use.

The laboratory fume hood has a baffle across the back that helps control the pattern of the air moving into and through the fume hood. The baffle is usually adjustable so the air flow through the hood can be directed up for lighter-than-air fumes, or down for heavier-than-air fumes. The baffle is normally constructed so it is impossible, by routine adjustment, to restrict air flow through the fume hood more than 20%.

The top panel of the fume hood has an exhaust collar to connect the exhaust duct to the fume hood. The exhaust duct may have a manual or automatic damper to control total volume of air through the hood.

Total air volume may also be controlled by changing the speed of the exhaust fan or by moving volume dampers at the exhaust fan.

Most fume hoods have an airfoil, called a deflector vane, at the entrance to the work surface. The design of the vane smooths the flow of air across the work surface and deflects it to the lower baffle opening. When a deflector vane is installed there is a fixed opening between the work surface and the vane. Therefore, when the exhaust fan is on, there is always air flow into the hood even when the vertical sash is fully closed. Laboratory fume hoods are generally grouped into three basic classifications: conventional, auxiliary, and bypass.

Constant Air Volume Conventional Fume Hood

The conventional fume hood has a movable, vertical or combination horizontal-vertical sash. As the sash is lowered in this type of hood the face area is reduced and the velocity of the air through the face opening begins to increase to try to maintain a constant air volume. However, at some point in the closing of the sash the total volume of air is also reduced even with the increase in face velocity. Closing the sash on a conventional hood in a constant air volume system disrupts the air flow pattern. At some point in the closing of the sash high velocities and unwanted turbulence are created at the hood face. These high velocities and air turbulence can induce contaminants out of the fume hood into the laboratory space.

Constant Air Volume Auxiliary Air Fume Hood

The auxiliary air fume hood has a movable, vertical or combination horizontal-vertical sash with auxiliary air ducted into the hood. With the sash open, the auxiliary air is distributed across the face area before its passage into the hood. With the sash closed, the auxiliary air is introduced directly into the fume hood interior.

Auxiliary air fume hoods are designed to reduce the amount of conditioned laboratory air required for make-up of exhaust air. Some hoods are specified to have 70% auxiliary air and 30% room air. In theory, this makes this hood more energy efficient because the auxiliary air is only partially heated or cooled. However, tests have shown that the best safety performance occurs when the auxiliary air is slightly warmer than the laboratory room air. Other concerns about this type of fume hood are that the auxiliary air entering the hood may create turbulence that can cause the air to reverse flow back out the hood face. Also, unconditioned auxiliary air may not enter the hood properly due to changes in density that occur with changes in outside air temperature.

Constant Air Volume Bypass Fume Hood

The standard bypass hood has a movable, vertical or combination horizontal-vertical sash. The construction of the bypass hood is similar to the standard conventional hood described above with the addition of the bypass. As the sash is pulled down, the air volume through the hood face is reduced. Simultaneously, as the sash is being closed, the bypass is being opened, and more air is then drawn through the bypass. This keeps the total air flow through the hood and the face velocities relatively constant. This is a major improvement over the standard constant air volume conventional fume hood.

Variable Air Volume Conventional Fume Hood

The air flow through the variable air volume conventional fume hood is variable in total air volume but has a constant face velocity. This hood has a movable, vertical or combination horizontal-vertical sash and is equipped with special controls to allow the volume of exhaust air to vary while still maintaining a constant velocity across the hood face. As the sash is lowered, the face area is reduced. The face velocity begins to increase to maintain constant volume as described with the standard constant volume conventional hood. However, unlike the standard constant volume conventional hood, as the velocity of the air through the face increases a controller in the hood or sash senses the rise in velocity and sends a signal to an air valve in the exhaust duct or to the exhaust fan to decrease air volume through the hood to maintain a pre-selected face velocity. When the sash is raised, the controller senses a face velocity below the set point and sends a signal to increase air volume to maintain the correct face velocity. The variable air volume fume hood and exhaust system reduce the volume of conditioned air exhausted as the sash is closed which results in energy savings.

TESTING AND BALANCING LABORATORY FUME HOODS

Tests to determine the operating condition of a fume hood must be conducted on new systems and then periodically as required by local code. Performance tests are conducted for volume of flow, face velocity and reverse air flow. The tests give a relative and quantitative determination of the efficiency of the fume hood. The test procedure outlined here presumes a bench-type laboratory fume hood in an air conditioned environment. If other types of hoods are used, some modification of the test procedure may be required.

Instruments and Equipment

The following list is typical of the test instruments and equipment needed to perform the various air flow and smoke tests. Exercise care when using chemical smoke.

Air Flow
Manometers
     liquid filled or electronic micro manometer
Pitot tube
Static pressure tip
Velocity grid
Electronic anemometer

Smoke
Smoke producing device
     candle, 1/2 minute
     tubes
     titanium tetrachloride
     dry ice (solid CO_2)

Cotton swabs, or a small brush and masking or duct tape for the titanium tetrachloride.
Container, warm water and tongs or gloves for dry ice.

Face Velocities

Generally, the toxicity level of the work done within the hood will determine the hood face velocities and the total amount of the exhaust air. Materials of little or no toxicity only need face velocities sufficient to maintain control under normal operating conditions. As toxicity levels rise, the face velocity should be increased to assure control. The recommended velocities for different toxic levels range from 80 fpm to 150 fpm. Generally, 100 fpm is satisfactory for most applications. The design engineer, fume hood user, industrial hygienist, or the applications engineer must specify what face velocity is required for each hood. The velocities must agree with the toxicity level of the work to be performed in the fume hood and the safety standards established by the facility.

Test Conditions

Building and laboratory air conditioning systems should be operating normally. General activity in the laboratory should be normal. Air currents in front of the hood should be reduced or eliminated if possible. The velocity of the air currents in front of the hood should never exceed 20% of the average required face velocity. Conduct the test with the normal hood apparatus in place and operating except where clearance must be provided for the test instruments or other equipment. Set a vertical-moving sash in the full open or in the "as used" position. For a combination sash, close the vertical sash and position the horizontal sash to get the maximum face opening or set it in the "as used" position. Make a sketch of the room showing the general layout of the laboratory. Include the location of all fume hoods and other significant laboratory equipment. Also include on the sketch the layout of the supply, return, and exhaust air duct and air devices.

Sash Test Procedure

Test the mobility of the sash. Grip the sash on the right side and raise and lower the sash. Repeat on the left side. The sash should glide smoothly and freely and hold at any height without creeping. Record the findings on the test sheet.

Air Volume and Velocity Test Procedure

Where practical, traverse the exhaust duct using a manometer to determine total volume of air through the hood. Calculate the total air volume, in cubic feet per minute, by multiplying the area of the duct by the average duct velocity $(Q = AV)$. Take static pressures in the exhaust duct at the collar and compare with design static pressure loss through the fume hood. Record the readings on the test sheet. All holes drilled for testing must be sealed as specified by the engineer or hood manufacturer.

Use a velocity grid to traverse the face opening. If a velocity grid is not available divide the hood face into a grid using the vertical and horizontal dimensions to get equal areas over the cross section of the opening. Establish the center of each area. The maximum distance between the centers should not exceed 12 inches. Make a sketch showing area centers. Traverse the opening using an electronic anemometer and take a velocity reading at each center point. The hood fails the test if the minimum reading at any point on the traverse is less than 80% of the average face velocity. Record the readings on the test sheet. Calculate the average face velocity in feet per minute. Compare the average face velocity with design specifications. Calculate volume of exhaust air at the hood face. Multiply the square feet of the opening by the calculated average face velocity $(Q = AV)$. Record the volume on the test sheet. Compare with the air volume taken at the exhaust duct traverse point. As applicable, place a certificate showing test results on the fume hood.

Smoke Test Procedure

Make a complete traverse of the hood face with either titanium tetrachloride on a cotton swab or brush, a smoke candle, a smoke tube, or other appropriate device to determine that a positive air flow is entering the hood over the entire opening. Use a smoke tube, a smoke candle, or brush a stripe of titanium tetrachloride on pieces of tape to produce the necessary amount of smoke at the following test locations: Along both sides, the top and the work surfaces of the hood about 6 inches behind and in parallel to the hood face. Along the back panel and the baffle of the hood. Around any equipment in the hood.

Verify that all smoke is carried to the back of the hood and exhausted. The hood fails the test if visible smoke flows out the front of the hood. Reverse air flows or dead air spaces are not permitted.

With the sash open, ignite a smoke candle within the hood enclosure to observe the exhaust capacity of the hood. All smoke should flow quickly and directly to the back of the hood and be exhausted. Set the candles on the work surface and close the sash. With the sash closed, the hood must have enough air to dilute and exhaust the smoke. The hood fails the test if visible smoke flows out the front of the hood. Reverse air flows or dead air spaces are not permitted.

Place a container of warm water in the center of the work area and add enough chunks of dry ice to the water to form a large volume of heavy white smoke. All smoke should flow directly to the back of the hood and be exhausted. The hood fails the test if visible smoke flows out the front of the hood. Reverse air flows or dead air spaces are not permitted.

TROUBLESHOOTING

Reverse Air Flow

Reverse air flow in a fume hood can be caused by air currents in the room or eddy currents in the hood. Adverse room air currents can be produced by people walking past the hood, by opening and closing doors to the laboratory, and by the air velocity and air pattern from the supply outlets. To reduce the influence from room air currents, fume hoods should be at least 6 feet away from doors and active aisles. Also, supply air devices should be selected and installed to avoid air velocities and patterns that would adversely affect the performance of the hood. Several small ceiling diffusers should be used instead of one large diffuser for each ceiling area. Perforated duct or perforated ceilings are other supply air options. Supply air outlets should be at least 4 feet from the hood face and terminal velocity should be 50 fpm or less. Sidewall outlets should be installed on adjacent walls instead of the wall opposite the fume hood face or set to discharge air over the top of the fume hood. Air flow from supply air outlets discharging directly into the fume hood should be reduced to acceptable levels or the pattern deflectors in outlets should be set to discharge away from the hood.

Eddy currents in the hood are caused by depressions, sinks, posts, service fittings and sharp edges. A plain, sharp edge at the entrance of the fume hood, for example, will produce turbulence about one inch above the work surface and about 6 inches into the hood. Fumes generated in this area will be disturbed and could possibly escape the hood. Eddy currents can also be created by blenders. mixers, and other equipment and instrumentation in the hood. To reduce eddy currents, airfoils should be installed on hoods with a plain sharp-edged entrance. Sinks, service fittings, etc., should be placed toward the back of the hood and all research operations should be conducted at least 6 inches into the hood.

Reverse air flow in a laboratory hood can be caused by the movements of the user at the hood face. Do a smoke test of the hood while simulating user movement at the hood face. Correct any conditions causing reverse flows. Reverse air flow in a laboratory hood can also be caused by high face velocities. Reduce face velocities to 150 fpm or less.

Face Velocities

Changes in the size of the face opening, the position of the baffle, the position of the volume damper, the ductwork, the exhaust fan speed, the filters, or blockages of the air path all affect hood face velocity. Review the following methods to make a permanent change in the hood face velocity. If changes are made, take new air pressure and electrical power measurements on the exhaust fan motor. To reduce face velocities consider using any of the following appropriate methods.

•   Slow the fan speed by decreasing motor sheave pitch diameter or increasing fan sheave pitch diameter. Change belts if needed.

•   If the system is VAV and has a speed or static pressure controller, set the controller for a slower speed or a lower static pressure.

•   Close the volume damper.

•   If the system is VAV, reduce the air volume at the air valve.

•   Open the sash on a constant volume conventional hood. If the hood is a constant volume bypass type or the system is VAV, the face velocity should not change. If face velocity does change, investigate and correct where practical.

To increase face velocities consider using any of the following appropriate methods.

•   Increase the fan speed by increasing motor sheave diameter or decreasing fan sheave pitch diameter. Change belts if needed.

•   If the system is VAV and has a speed or static pressure controller, set the controller for a faster speed or a higher static pressure.

•   Open the volume damper.

•   If the system is VAV, increase the air volume at the air valve.

•   Close the sash on a constant volume conventional hood. If the hood is a constant volume bypass type or the system is VAV, the face velocity should not change. If face velocity does change, investigate and correct where practical.

•   Remove any blockages in the air path. Visually inspect the baffle openings and behind the baffle. Take static pressures in the duct and across the filters and the fan. Visually inspect where possible. Where practical, change ductwork to relieve restrictions and reduce duct friction and dynamic losses.

TERMINOLOGY

Access Opening: That part of the fume hood through which work is performed.

Airfoil: The curved vane along the bottom of the hood face at the fume hood entrance. Used to counteract the effect of the vena contracta generated at the opening of plain entrance hoods. See deflector vane.

Air Lock: An anteroom with airtight doors between a controlled and uncontrolled space.

Auxiliary Air: Supply or supplemental air delivered to a laboratory fume hood to reduce room air consumption. Exhaust air requirements for laboratory fume hoods and other containment devices often exceed the supply air needed for the normal room air conditioning. To meet these needs auxiliary air can be ducted directly into the fume hood or supplied to the room. The auxiliary air should be conditioned to meet the temperature and humidity requirements of the lab. Energy savings may be obtained when heating, cooling and humidifying requirements are kept to a minimum. In some laboratory buildings a corridor separates the offices from the labs. The offices are under positive pressure. Air leaves the offices and enters into the corridor. The corridor acts as a plenum. The air then flows into the negatively pressurized lab. The result is that it may be possible to reduce the amount of auxiliary air.

Baffle: A panel located across the back of fume hood that controls the pattern of air moving into and through the fume hood.

Bench Mounted Hood: A fume hood that rests on a counter top.

Biological Safety Cabinet: A special safety enclosure that uses air currents to protect the user. Used to handle pathogenic microorganisms. Biological safety cabinets are also called safety cabinets, laminar flow cabinets, and glove boxes.

Bypass: A compensating opening that maintains a relatively constant volume of exhaust through a fume hood, regardless of sash position. The bypass functions to limit the maximum face velocity as the sash is closed.

California Hood: A rectangular enclosure used to house distillation and other large research apparatus. It can provide visibility from all sides with horizontal sliding access doors along the length of the assembly. The California hood is not considered a laboratory fume hood.

Canopy Hood: A suspended ventilating device used to exhaust only heat, water vapor and odors. The canopy hood is not considered a laboratory fume hood.

Capture Velocity: The air velocity at the hood face necessary to overcome opposing air currents and to contain contaminated air within the laboratory fume hood.

Cross Draft: A flow of air that blows across or into the face of the hood. Cross drafts, created by the room ventilation system, a corridor or from an open door, if located adjacent to the fume hood can drastically disturb the flow of air into the hood face and even cause reverse flow of air out the front of the hood.

Dead Air Space: Lack of air movement in the hood.

Deflector Vane: An airfoil-shaped vane along the bottom of the hood face that deflects incoming air across the work surface to the lower baffle opening. The opening between the work surface and the deflector vane is open even with the sash fully closed.

Diversity Factor: Used with a variable air volume system. A diversity permits the exhaust system to have less capacity than that required for the full operation of all units.

Effluent: Outflow

Exhaust Collar: The connection between the duct and the fume hood through which all exhaust air passes.

Face Velocity: The average velocity of air moving into the fume hood opening. Expressed in feet per minute (fpm).

Fume Hood Exhaust System: An arrangement consisting of a laboratory fume hood, ductwork, fan, and all other equipment and controls required to make system operable.

Glove Box: An enclosure used to confine and contain hazardous materials with user access by means of gloved portals or other limited openings. Glove boxes provide greater protection but are more restrictive than laboratory fume hoods or other biological safety cabinets. Glove boxes require far less exhaust air than laboratory fume hoods or other biological safety cabinets.

High Efficiency Particulate Air (HEPA) Filters: A filter with an efficiency in excess of 99.97% for 0.3 micrometer particles as determined by Dioctyl Phthalate (DOP) test.

Hood Face: The plane of minimum area at the front portion of the hood through which air enters when the sash is fully open.

Infiltrated Air: Auxiliary air induced from the corridor or other spaces into the lab.

Laminar Flow Cabinet: A clean bench or biological safety cabinet that uses smooth directional air flow to capture and carry away airborne particles. The laminar flow cabinet is not considered a laboratory fume hood.

Liner: Interior lining of a laboratory fume hood used for the side, back and top enclosure panels, exhaust plenum and baffles.

Lintel: The portion of a laboratory fume hood located directly above the access opening.

Particulate: Of, pertaining to, or formed of separate particles.

Perchloric Acid Hood: A special purpose hood designed primarily to be used with perchloric acid. Perchloric acid fume hoods are mandatory for research in which perchloric acid is used because of the explosion hazard associated with this chemical.

Perforated Ceiling: An air distribution device. Perforated ceiling panels or filter pads are used to distribute the air uniformly throughout the ceiling or a portion of the ceiling.

Perforated Duct: An air distribution device. Perforated ducts are used to distribute the air uniformly throughout the laboratory space.

Plenum: An air chamber or compartment.

Radioisotope Hood: A special purpose hood designed primarily for use with radiochemicals or radioactive isotopes. Special filters and shielding are required. Radioisotope hoods have exhaust ducts with flanged, neoprene gaskets with quick disconnect fasteners that can be quickly dismantled for decontamination.

Reverse Air Flow: Air movement toward the front of the hood.

Sash: The moveable, normally transparent panel set in the fume hood entrance. Sashes may be vertical or a combination horizontal and vertical. The combination horizontal/vertical sash has horizontally sliding sashes set in a vertical rising sash. With the combination sash the vertical sash allows for easier setup or removal of hood equipment or apparatus while the horizontal sash facilitates user operations and reduces total exhaust air volume.

Slot Velocity: The speed of the air moving through the fume hood baffle openings.

Smoke Candle or Tube: A smoke producing device used to allow visual observation of air flow.

Specified Rating: The hood performance rating as specified, proposed, or guaranteed.

Spot Collector Hood: A small, localized ventilation hood usually connected by a flexible duct to the exhaust system. The spot collector hood is not considered a laboratory fume hood.

Threshold Limit Values (TLV): The values for airborne toxic materials that are to be used as guides in the control of health hazards and represent time-weighted concentrations to which nearly all workers may be exposed 8 hours a day over extended periods of time without adverse effects.

Titanium Tetrachloride (TiCl4): Titanium tetrachloride is a chemical that generates white smoke. It is used to test the air pattern in laboratory fume hoods. Titanium tetrachloride is corrosive and irritating. It can stain the hood and will produce a residue that must be cleaned up. Care must be taken to minimize the effects on the hood. Also avoid skin contact or inhalation.

Zone Pressurization: Zone pressurization is a means of isolating spaces that generate harmful contaminants. Zone pressurization means that the air distribution system is designed so the hazardous areas have negative pressure and any airborne contaminants are contained in the negatively pressurized areas.

PART 2

Chapter 11
Basic Water Balance Procedures

This chapter describes procedures for balancing water systems using direct flow measurements. This is the preferred method because it eliminates the compounding errors introduced by balancing using pressure drops across components or using temperature measurements. Unfortunately, a great many systems don't have flow meters. Therefore, this chapter will also address balancing using pressure drops across system components and as a last resort, balancing with temperatures.

BALANCING PRINCIPLES

Balancing is measuring and adjusting the system to get the design water flow. Unless specifications require otherwise, it's normally considered there's an adequate balance when the water quantities measured are within plus or minus 10 percent of design quantities.

General Balancing Procedure
1. Set the controls for full flow through the coils.

2. Set the pressure reducing valve (PRV) to maintain the proper pressure in the system. The PRV should be adjusted so there's a minimum of 5 psi additional pressure at the highest terminal.

3. Set all manual valves full open.

4. Proportionally balance the system.

5. If necessary, adjust the impeller diameter to bring the system to within 10% of design flow.

6. If the system has three-way valves it has been balanced with the valves open to full flow through the coil and closed to flow

through the bypass. Change the controls so there's full flow through the bypass and no flow through the coil. Determine flow, either by flow meter or pressure drop across the valve, and adjust the balancing valve in the bypass piping so full flow gpm through the bypass is the same as was full flow gpm through the coil. In most cases, the bypass balancing valve should be throttled since normally the coil has a higher pressure drop than the three-way valve and more water will flow through the valve when it's in the bypass position. If the bypass balancing valve is not adjusted, more water than is required will also flow through the bypass when the valve is in a modulated position. The coil will receive less than design gpm. For example, if the valve is modulated at 50% bypass and 50% flow through the coil and the bypass balancing valve has not been set the coil may receive only 25% of flow while the bypass gets 75%.

7. Complete report.

PROPORTIONAL BALANCING
USING DIRECT MEASUREMENT FLOW METERS
Determining Total Flow

Determine the total volume using a manometer and Pitot tube traverse of the main, or a reading of the main flow meter using a differential pressure gage or a Bourdon tube test gage. For approximately total flow, read the pump.

If the flow isn't within 10% of design, try to determine the reasons for the difference. If it's determined that everything is in order but the total flow is greater than 20% above design, close the pump discharge valve or adjust the main balancing valve to bring the water volume down to 10 to 15% above design. If, however, the total flow is below design, confirm that the discharge valve and the main balancing valve are full open. Low flow problems may also be caused by poor inlet or outlet connections at the pump and restrictive fittings in the piping system. Consult with the mechanical contractor if any changes are thought necessary. If the flow is still below design, a decision must be made to determine if the system will be proportionally balanced low or if new impeller is needed. Use the pump laws to calculate the new impeller size needed to bring the water flow as close to 110% of design as possible. It's important to try to set the flow between 100% and 110% since during the balancing process there will be some loss of total water; generally between 5 and 10%.

It's possible that by increasing the impeller size a motor or pump change may also be needed. Therefore, after calculating the impeller size the new required brake horsepower must be calculated to determine if there's adequate horsepower presently available. Never increase the pump flow to a point where the motor is in an overloaded condition.

Proportionally Balancing The Distribution System

Proportionally balancing the water distribution system is testing and adjusting the water flow at each terminal first and then at each branch, riser and lastly, the header. If the system has a primary and secondary loop then the primary circuit is balanced first.

To reduce flow, balancing valves in the system will be adjusted. When a valve is closed the static head upstream of the valve is increased. In theory, this means that the pump is working against a greater static head and will mean a decrease in total water flow. Therefore, when several valves are throttled it can be expected there will be an increase in static head at the discharge of the pump and a reduction in total flow. This is why when starting the balance it's good practice, if possible, to set the pump at 110% of design. Even with this cushion it still may be necessary to open the discharge valve or main balancing valve (if it had been throttled) or increase the impeller size when the balancing is finished to bring the water flow to design conditions.

The principles of proportional balancing require that to begin the balance all the valves in the system must be full open and at least one terminal balancing valve (the terminal with the lowest percent of design flow) will be open when the balance is finished. If the system has branches, risers and headers at least one balancing valve in each (the one with the lowest percent of design flow) will also be full open when the balance is finished.

Procedure:

1. Read the entire system. If a part of the distribution system is extremely low and the problem can't be corrected before starting the balancing, for instance, because the problem is poor

piping design, don't sacrifice the entire system by cutting the major portion of the system in an attempt to force water to the problem area. Simply, ignore the problem area and proportionally balance the rest of the system. If the balance can be delayed, consult with the design engineer for solutions to correct the problem. For example (1) redesign restrictive piping, (2) design and install a new separate secondary system.

2. Determine which terminal has the lowest percent of design flow. Design flow is either the original flow per the contract specifications or a new calculated design flow. Percent of design flow is equal to the measured flow divided by the design flow:

$$\%D = \frac{M}{D}$$

Typically this will be on the branch on the riser farthest from the pump.

3. Starting with the terminal with the lowest percent design flow, as needed, adjust each terminal on that branch in sequence, from the lowest percent of design flow to the highest percent of design flow so the ratio of the percent of design flow (ratio of $\%D$ = terminal X $\%D$/terminal Y $\%D$) between each terminal is plus or minus 10%.

4. Go to the branch on this riser that has the terminal with the next lowest percent of design flow as determined from the initial readings. Typically, this terminal will be on the second farthest branch. Balance all the terminals on this branch to each other within plus or minus 10% of design.

5. Continue balancing the terminals on every branch in the system to within plus or minus 10% of each other.

6. Starting with the branch in the system with the lowest percent of design flow proportionately balance all branches on each riser from the lowest percent of design flow to the highest percent of design flow to within 10% of each other.

7. When all the branches on all the risers have been proportionally balanced start with the riser that has the lowest percent of design flow and proportionally balance each riser.

8. After the risers have been balanced, proportionally balance the headers.
9. If necessary, adjust the impeller diameter to bring the system to within 10% of design flow.
10. Read the entire system.
    a. The balancing valve on one terminal on each branch should be full open.
    b. The balancing valve on one branch on each riser should be full open.
    c. The balancing valve on one riser on each header should be full open.
    d. The balancing valve on one header should be full open.
11. Make any final adjustments. If the system has three-way valves follow the procedure outlined in the general balancing procedures.

Example 11.1: Balance the chilled system illustrated in figure 11.1.
1. All the coils have been read and coil No. 1 on branch No. 1, riser No. 1 has the lowest percent design gpm.

Coil	%D	Ratio
1	78	2:1 = 1.12
2	87	
3	100	
4	122	
5	125	

2. Starting with coil No. 1 adjust each coil on branch No. 1 until the ratio of the percent of design flow between each coil is plus or minus 10%.
    a. Adjust coil No. 2 to coil No. 1.
    b. Adjust coil No. 3 to coil No. 2.
    c. Adjust coil No. 4 to coil No. 3.
    d. Adjust coil No. 5 to coil No. 4.
3. Go to the branch No. 2 on riser No. 1. Starting with the coil with the lowest percent of design gpm, balance all the coils on branch No. 2 to within plus or minus 10% of desired flow.
4. Continue until all the coils on both branches of riser No. 1 have been balanced to within plus or minus 10% of each other.

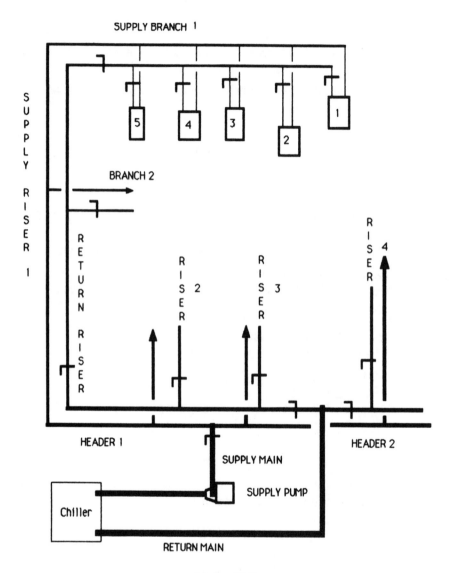

Figure 11.1

5. Starting with branch No. 1 which has the lowest percent of design flow, proportionately balance No. 2 to branch No. 1.

Branch	%D	Ratio
1	89	2:1 = 1.13
2	101	

6. After the branches on riser No. 1 have been proportionally balanced, go to the next riser and start with its lowest branch. When all the branches on all the risers have been proportionally balanced start with the riser that has the lowest percent of design flow and proportionally balance each riser.

Header	Riser	%D	Ratio
1	1	95	2:1 = 1.12
	2	106	
	3	120	
2	4	138	
	5	122	
	6	110	

Balance:
a. Riser No. 2 to riser No. 1.
b. Riser No. 3 to riser No. 2.
c. Riser No. 5 to riser No. 6.
d. Riser No. 4 to riser No. 5.

7. After the risers have been balanced, proportionally balance header No. 2 to header No. 1.

Header	%D	Ratio
1	105	2:1 = 1.16
2	122	

8. If needed, after balancing header No. 2 to header No. 1, adjust the impeller diameter to bring the system to within 10% of design flow.

9. Read the entire system and make any final adjustments.

PROPORTIONAL BALANCING USING DIFFERENTIAL PRESSURE

The use of direct reading flow meters is considered the most reliable and accurate method of determining flow in a water system.

However, since flow meters aren't always installed, measuring the pressure drop across a terminal coil or other primary heat exchanger using a differential pressure gage or a Bourdon tube test gage can also produce total flow quantities and a means for balancing flow.

When using a Bourdon tube gage ensure that the gage is at the same height for both the entering and leaving reading or that a correction is made for a difference in height. To eliminate the need for correction and the possibility of error resulting from the correction, it's best to use one test gage hooked up with a manifold. With a single gage connected in this manner, the gage is alternately valved to the high pressure side and then the low pressure side to determine differential, eliminating any problem about gage elevation.

Balancing Procedure

Use the balancing procedures as outlined in this chapter and proportionally balance the terminals using flows determined from pressure drop across the terminals or valves.

Measuring Flow Across a Terminal

To use this method of flow measurement the coil or heat exchanger must:

1. Be a new coil or in new condition. The coil is acting like an orifice plate or any other flow meter with a known pressure drop. The tubing in the coils have a known resistance to a certain flow rate. If the flow rate is increased the resistance to flow is also increased, by the square of the flow. This can only happen if the coil is new. When the rated flow and pressure drop are known use the following equation to determine actual flow:

Equation 11.1

$$GPM_2 = GPM_1 \sqrt{\frac{PD_2}{PD_1}}$$

Where:
GPM_1 = rated flow in gallons per minute
GPM_2 = actual flow in gallons per minute
PD_1 = rated pressure drop
PD_2 = actual pressure drop

If the coil is old and scaled up, the inside diameter of the pipe is reduced and the resistance is increased at rated flow. The original rated pressure drop for rated flow is no longer valid and equation 11.1 can't be used.

2. Have certified data from the equipment manufacturer showing actual tested water flow and pressure drop at rated flow. If the data was derived by calculation instead of from actual tests, the information may not be accurate.

3. Have properly located pressure taps for measuring differential pressure. Some installations have the pressure taps at poor locations that may lead to erroneous readings because of added losses from piping and fittings.

Example 11.2: A terminal is rated at 30 gpm with 5.6 feet of pressure drop. The entering pressure is measured at 28 feet and the leaving pressure is 24 feet. Find the actual flow rate.

Answer: 25.4 gpm

Solution:

$$GPM_2 = GPM_1 \sqrt{\frac{PD_2}{PD_1}}$$

$$GPM_2 = 30 \sqrt{\frac{4}{5.6}}$$

$$GPM_2 = 25.4$$

Measuring Flow Across A Valve

Another way to determine flow is to use the drop across a control valve. As with taking pressures across a terminal, the valve must be in new condition and the pressure taps must be installed as close to the inlet and outlet of the valve as possible. Control valve manufacturers rate their valves based on the relationship between pressure drop and flow. The term for this relationship is "flow coefficient" (C_V) and is the flow rate in gpm of 60-degree water which will cause a pressure drop of 1 psi (2.31') across a wide open valve. Where the C_V is known the flow in gpm may be determined by the following equation.

Equation 11.2

$$GPM = C_V \sqrt{PD}$$

Where:

GPM = water flow rate in gallons per minute
C_V = flow coefficient or valve constant
PD = measured pressure drop across the valve in psi or feet

Example 11.3: An automatic control valve has a C_V of 25. The valve is placed in the full open position and the entering pressure is measured at 5 psi. The leaving pressure is 3.5 psi. Find the flow through the valve.

Answer: 30.6 gpm

Solution:

GPM $= C_V \sqrt{PD}$

GPM $= 25 \sqrt{1.5}$

GPM $= 30.6$

Example 11.4: Fourteen gpm is flowing through a wide open valve with a C_V of 10. To bring the flow rate down to 12 gpm, the balancing valve is partially closed. To determine how far to close the balancing valve, the pressure drop across the control valve must be found. Use the following equation.

Equation 11.3

$$PD = \left(\frac{GPM}{C_V}\right)^2$$

Answer: 1.44 psi

Solution:

$$PD = \left(\frac{GPM}{C_V}\right)^2$$

$$PD = \left(\frac{12}{10}\right)^2$$

PD = 1.44

PROPORTIONAL BALANCING
USING DIFFERENTIAL TEMPERATURES

Temperature measurements should be used as a check and not a primary balancing procedure. Balancing by temperature measurement is unreliable because of varying load conditions, air stratification and errors in using thermometers and is not recommended as an accurate balancing procedure on chilled water systems or on hot water systems with (1) high design temperature differentials, (2) different type components, (3) components with different design temperature differentials. This leaves low temperature hot water heating systems in the range of 200 degrees with the same design temperature differentials of 40 degrees or less and using the same components.

To help prevent errors in measurement, check that temperature wells or other temperature measuring stations are properly installed at good locations, use the same thermometer for all readings, and allow ample time for the temperatures to stabilize. Although not a recommended procedure, pipe surface temperature measurements can be used for approximating the temperature of hot water heating systems. The surface temperature of water pipes should be above 150 degrees Fahrenheit and the pipe surface must be cleaned to a bright finish. Care must be exercised to avoid error caused by sensing ambient air temperature.

Temperature Balance Procedures Using
Differential Temperatures to Determine Coil GPM

1. Use the general balancing procedure as outlined in this chapter.
2. Proportionally balance the terminals using flows determined from temperature drop across the terminals.
   a. Coil gpm may be calculated when the actual heat transfer rate (the airflow has been balanced) in Btuh (British thermal units per hour) and the actual water temperatures are known.

Equation 11.4:

$$GPM = \frac{BTUH}{500 \times TD}$$

Where:
GPM   = flow rate through the coil in gallons per minute
For heating coils and chilled water coils with no latent load (dry coil):

BTUH = CFM x 1.08 x ΔT
(ΔT = temperature difference between entering and leaving air)
For chilled water coils with latent load (wet coil):
BTUH = CFM x 4.5 x Δh
(Δh = enthalpy difference, in Btu/lb, between entering and leaving air)
500 = constant
TD = temperature difference between entering and leaving water

Example 11.5: A hot water coil has an entering temperature of 180 degrees and a leaving temperature of 160 degrees. The heat transfer rate through the coil has been calculated at 60,000 Btuh. Find the flow rate.

Answer: 6 gpm

Solution:

$$GPM = \frac{BTUH}{500 \times TD}$$

$$GPM = \frac{60,000}{500 \times 20}$$

$$GPM = 6$$

b. Coil gpm may also be calculated when (1) the design gpm and temperature differential and (2) the measured temperature differential are known. This is assuming that the airside load is at design conditions (very unlikely).

Equation 11.5:

$$GPM_2 = GPM_1 \frac{\Delta Td}{\Delta Tm}$$

Where:
GPM_2 = Calculated gallons per minute
GPM_1 = Design gallons per minute
ΔTd = Design water temperature differential
ΔTm = Measured water temperature differential

Example 11.6: A coil is designed for 28 gpm. The design temperature differential is 20 degrees. The measured temperature differential is 25 degrees. Find coil gpm.

Answer: 22.4 gpm

Solution:

$$GPM_2 = GPM_1 \frac{\Delta Td}{\Delta Tm}$$

$$GPM_2 = 28 \frac{20}{25}$$

$$GPM_2 = 22.4$$

Temperature Balance Procedures
Using Temperature Differentials
1. Use the general balancing procedure as outlined in this chapter.
2. Determine the temperature differential for each coil.
3. Proportionally balance.

CONDITION I: Coils have the same design entering and leaving water temperatures. The air has been balanced and the airflow is constant. The load is constant.

1. Supply water temperature must stay constant during the balance but it's not necessary to have design temperatures.
2. Balance from the coil with the lowest return temperature to the coil with the highest return water temperature until the return water temperature at each coil is within 10% of each other.

Example 11.7: The design water temperatures are 180 degrees entering and 160 degrees leaving. The supply temperature is maintained at 175 degrees. Coil No. 1 has a leaving temperature of 165 degrees and coil No. 2 has a leaving temperature of 145 degrees.

1. The % of difference is 114% (165/145 = 1.14)
2. Cut the balancing valve for coil No. 1 and read coil No. 2 until the temperatures are within +/− 10%.

CONDITION II: Coils have the same design entering water temperatures and different leaving water temperatures. The air has been balanced and the airflow is constant. The load is constant.

1. Supply water temperature must stay constant during the balance

but it's not necessary to have design temperatures.

2. Balance from the coil with the lowest percent of design temperature differential to the coil with the highest percent of design temperature differential until the ratio percent of design temperature differentials is within +/- 10%.

Example 11.8: The design entering water temperature is 180 degrees. The supply temperature is maintained at 175 degrees. Coil No. 1 has a design leaving temperature of 160 degrees. Its measured leaving temperature is 165 degrees. Coil No. 2 had a design leaving temperature of 150 degrees. Its measured leaving temperature is 145 degrees.

Equation 11.6

$$\%D\ \Delta T = \frac{\Delta Td}{\Delta Tm}$$

Where:

$\%D\ \Delta T$ = Percent of design temperature difference
ΔTd = Design temperature difference
ΔTm = Measured temperature difference

Coil No. 1: Td = 20 (180−160)
Coil No. 1: Tm = 10 (175−165)
Coil No. 2: Td = 30 (180−150)
Coil No. 2: Tm = 30 (175−145)

$$\text{Coil 1 } \%D\ \Delta T = \frac{\Delta Td}{\Delta Tm}$$

$$\text{Coil 2 } \%D\ \Delta T = \frac{\Delta Td}{\Delta Tm}$$

$$\text{Ratio} = \frac{\text{Coil 1 } \%D\ \Delta T}{\text{Coil 2 } \%D\ \Delta T}$$

$$\text{Ratio} = \frac{\text{Coil 1 } \dfrac{\Delta Td}{\Delta Tm}}{\text{Coil 2 } \dfrac{\Delta Td}{\Delta Tm}}$$

$$\text{Ratio} = \frac{\dfrac{20}{10}}{\dfrac{30}{30}}$$

1. The ratio is 2 (200%/100%).
2. Cut the balancing valve for coil No. 1 and read coil No. 2 until the ratio is within 0.90 to 1.10.

Chapter 12
Centrifugal Pumps, Pump Performance Curves And Pipe System Curves

CENTRIFUGAL PUMPS

Centrifugal pumps are made in a wide variety of types. They're primarily classified according to type of casing (housing), impeller design, inlet, stages and whether they're direct coupled or belt driven.

Casing

The casing is usually a scroll or volute and is built of various metals, alloys and other materials depending on the liquid being pumped, and the temperature and pressure requirements of the liquid.

Impeller

Impellers are classified according to the configuration of the vanes and the angles at which the liquid enters and leaves the impeller. Classifications are (1) radial, (2) mixed, (3) axial. Vanes or blades are arranged in a circular pattern around an inlet opening ("eye" of the impeller) at the center of the impeller. In some pumps a diffuser, having a series of guide vanes surrounds the impeller.

Seals

The impeller is secured on a shaft mounted in suitable bearings and has a seal where it passes through the casing wall. The seal prevents water or air leakage. The two most commonly used sealing methods are the stuffing box and the mechanical seal.

Couplings

Except for some small pumps mounted directly on the extended motor shaft, pumps are usually connected to the motor through a shaft coupling. There are two basic types of pump couplings; the equalized spring and the flexible disc.

Shaft Alignment

Shafts should be aligned as closely as possible for quiet operation and the least coupling and bearing wear. Alignment of shafts is done by using a straightedge on the coupling, except if extreme accuracy is called for, then a dial indicator which measures alignment differences in thousandths of an inch is needed. The ideal alignment condition is when both shafts are in a straight line and concentric under all conditions of operation and shutdown. However, because of changes in liquid temperature and operating temperature of the pump and motor there will be minor changes in alignment.

PUMP OPERATION

As the pump impeller is rotated, centrifugal force causes the water in the pump to be thrown outward from the impeller into the casing. This outward flow through the impeller reduces the pressure at the inlet of the impeller, allowing more water to be forced in through the pump suction opening by atmospheric or external pressure. As the water leaves the impeller at a high velocity it's collected in the casing where the velocity is reduced and converted into pressure. The pressure (head) developed by a centrifugal pump depends on the impeller diameter and speed of rotation and is entirely the result of the velocity imparted to the water by the impeller.

MULTIPLE PUMP ARRANGEMENTS

Pumps in Parallel

When two or more pumps operate in parallel, the capacities and horsepowers are additive at equivalent heads. When two identical pumps operate in parallel the head remains constant and the gpm and the horsepower double.

Example 12.1: A single pump operating at 50 gpm @ 80 feet of head and 4 bhp is put in parallel with an identical pump. The result is a total flow of 100 gpm, requiring 8 bhp and operating at 80 feet of head. Each pump supplies one-half the required flow at the required head.

Pumps in Series

When two or more pumps operate in series the heads and horse-powers are additive at equivalent capacities. When two identical pumps operate in series the gpm remains constant and the head and the horsepower double. Pumps operating in series work essentially the same as a multi-stage pump.

Example 12.2: A single pump operating at 50 gpm @ 80 feet of head and 4 bhp is put in series with an identical pump. The result is a total flow of 50 gpm, operating at 160 feet of head and requiring 8 bhp. Each pump supplies the required flow at one-half the required head.

Combination Series and Parallel Operation

In some installations, a combination of both series and parallel pumping is used. An advantage of parallel, series or combination pumping is that it can reduce both first and long-term energy costs by using a combination of smaller line-mounted pumps instead of a single large base-mounted pump.

PUMP PERFORMANCE CURVES

A pump performance curve is a graphic representation of the performance of a pump. The performance curve relates flow (in gallons per minute) to pressure (in feet of head). Performance curves are also known as pump curves, characteristic curves, or head-capacity curves. They're developed by the manufacturer under carefully controlled test conditions. The "feet of head vs gpm" relation (fig. 12.1) is most often used because of the physical characteristics of centrifugal pumps and because it gives a general description of pump operation without being affected by water temperature or density. Some curves, however, show pressure in psi vs gpm or psi vs flow in pounds per hour. These are specific, not general curves and are related to a specific water temperature and density.

Pump curves are very similar to fan curves with the pressure scale being plotted vertically on the left as the ordinate and the flow scale plotted along the bottom as the abscissa. One difference is that the pressure-capacity curves for fans are referenced to fan speed, but since most pumps are direct connected to their motors and the pump speed stays constant, the head-capacity curves for pumps are

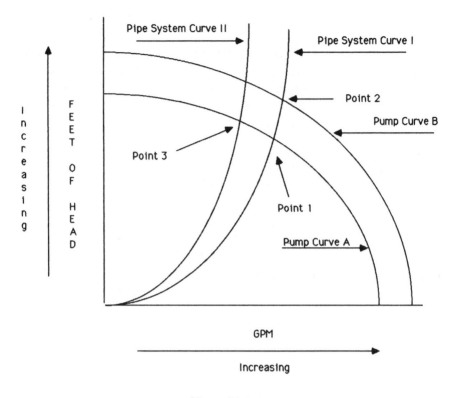

Figure 12.1

referenced to impeller size. As discussed in Chapter 9, to change the fan's capacity the discharge damper is open or closed or the fan speed is changed. For a change in pump capacity, the discharge valve is opened or closed or the pump impeller size is changed.

Although performance curves can be useful in troubleshooting pumps, you should be aware that because of installation conditions pump performance as determined by field tests is usually less than indicated by manufacturers' tests.

Composite Pump Curves

Besides the head-capacity curve, composite pump curves also give pump efficiency, brake horsepower, impeller size, NPSHR, model number, curve number, speed, inlet size, discharge size, maximum impeller diameter and minimum impeller diameter.

PIPE SYSTEM CURVES

System curves provide a means to analyze pump operation and to identify problem areas associated with losses in the system resulting from friction in the pipe, pressure drops through equipment, pressure loses in valves and fittings and, in open systems, the difference in static head. System curves are a plot of the change in head resulting from a change in gpm in a fixed piping circuit. The head changes as the square of the gpm (pump law No. 2).

Fixed Systems

A fixed system is one in which there are no changes in the system resistance because of closing or opening of valves, or changes in the condition of the coils, etc. For a fixed system, an increase or decrease in system resistance results only from an increase or decrease in gpm and this change in resistance will fall along the system curve. If however, valves are operated toward their closed positions, coils become corroded, etc., the system curve will no longer apply.

Closed System Curves

To plot a closed system (fig. 12.2) piping curve, use field measured gpm and head for the first point and then mathematically calculate (pump law No. 2) the other points.

Open System Curves

In open systems, besides the friction losses in the piping, the static heads must also be analyzed.

1. Suction static head is less than discharge static head (fig. 12.3). To plot the system curve, total static head must be added to the pipe friction loss head.

2. Suction static lift plus discharge static head (fig. 12.4). To plot the system curve, total static head must be added to the pipe friction loss head.

3. Suction static head is greater than discharge static head (fig. 12.5). To plot the system curve, total static head must be subtracted from the pipe friction loss head.

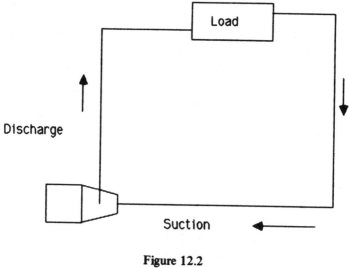

Figure 12.2
CLOSED SYSTEM

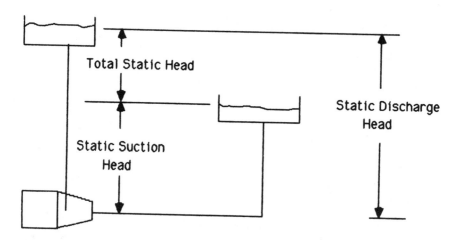

Figure 12.3
OPEN SYSTEM – SUCTION HEAD
IS LESS THAN DISCHARGE STATIC HEAD

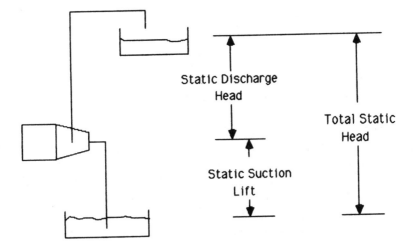

Figure 12.4
OPEN SYSTEM – SUCTION STATIC LIFT
PLUS DISCHARGE STATIC HEAD

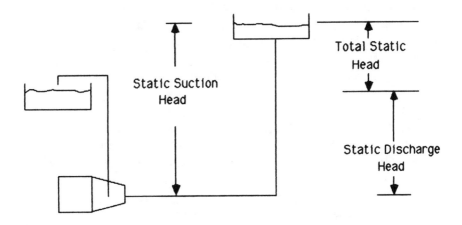

Figure 12.5
OPEN SYSTEM – SUCTION STATIC HEAD
IS GREATER THAN DISCHARGE STATIC HEAD

OPERATING POINT

Each piping system has its own system curve and each pump has its own performance curve. The intersection of these two curves is the operating point for both the piping system and the pump. Just as every system operates only along its system curve; every pump operates only along its performance curve. For the piping system, the head increases or decreases as the square of the gpm increases or decreases (pump law No. 2). However, for the pump, a decrease in head will mean an increase in gpm and vice versa.

When testing a pump, if it's found that the operating head exceeds the design head the measured gpm will be less than design. The horsepower needed is also less. Look for restrictions in the piping such as valves closed, poor inlet or discharge fittings, clogged strainers, etc.

If the operating head is lower than design the measured gpm will be greater than design. The horsepower and net positive suction head required (NPSHR) increases. This may result in an overloaded motor or pump cavitation. If this is the case, partially close the discharge valve to shift the head back to design. For extreme overpumping reduce the size of the pump impeller.

Using the pump laws, the pump performance curve, and the piping system curve, any change to the pump impeller or the piping system can be calculated and graphically depicted. For example, in figure 12.1 the operating point for a pump operating on pump curve A in pining system I is point 1. To increase or decrease gpm, a physical change must be made to either the piping system or the pump impeller. If the change is to the pump impeller, the pump will operate on a new performance curve that runs parallel to the original curve. Since the piping system has remained unchanged, the system curve (I) also remains unchanged. In figure 12.1, an increase in impeller size results in a new pump curve (B). The pump and the system are now operating at a higher gpm and head at point 2.

If, however, the increase or decrease in water flow is made by changing the piping system, by reducing or adding system resistance (for example, opening or closing a main valve), a new system curve is established while the pump performance curve (A) stays unchanged. In figure 12.1, an increase in system resistance results in a new system curve (II). The pump and system are now operating at a lower gpm and a higher head at point 3.

MULTIPLE PUMP ARRANGEMENTS

Parallel Pumping Systems

A system curve (fig. 12.6) can be used to determine the flow rate that a single pump in parallel will deliver, and the flow rate of both pumps. The design operating point for both pumps (point 2) is on the paralleled pump curve. The point of operation of each pump (when both pumps are running) is on the single pump curve (point 1). When only one pump is running the point of operation shifts to the intersection of the single pump curve with the established system curve (point 3).

One of the benefits of parallel pumping is standby protection. The function of the standby pump is to take over or continue the pumping function when the regular pump goes out of service. You'll notice that with parallel pumping if one pump goes out of service the remaining pump can still deliver a high degree of design flow. In fact, the flow rate from the single pump (point 3) actually increases from the flow rate that it delivered as a single pump when both pumps were running (point 1).

Parallel pumping requires that each pump motor be sized for maximum horsepower. Because of the increase in flow rate, maximum horsepower will occur when only one pump is operating. This is the reverse of series pump operation.

Series Pumping Systems

A system curve (fig. 12.7) can be used to determine the flow rate that a single pump in series will deliver, and the flow rate of both pumps. Putting pumps in series steepens the overall pump curve. The design operating condition for both pumps is point 1. The point of operation of each pump (when both pumps are running) is on the single pump curve (point 2). When only one pump is running the point of operation shifts to the intersection of the single pump curve with the established system curve (point 3). Notice that when the system goes to single pump operation the head increases over the head of the single pump when both pumps are on. The flow delivered decreases. The horsepower required also decreases.

Series pumping requires that each pump motor be sized for maximum horsepower. This will happen when both pumps are operating (point 2). This is the reverse of parallel pump operation.

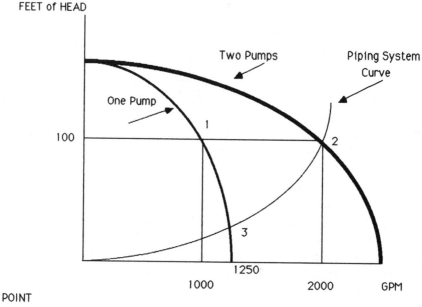

POINT
1
The actual point of operation of each pump when both pumps are on.
2
The operating point on the system curve when both pumps are on.
3
The operating point on the system curve if one pump is taken out of service.

Figure 12.6
PARALLEL PUMPS

NET POSITIVE SUCTION HEAD

The suction conditions of a centrifugal pump are among the essential factors affecting its operation. The centrifugal force of the pump's rotating action, plus impeller design and casing design make the pump's discharge pressure higher than the suction pressure. Therefore, a partial vacuum is created at the pump inlet and atmospheric or external pressures force water into the pump. After the water enters the pump's suction opening there's a further reduction of pressure between this opening and the inlet of the impeller. This is the lowest pressure in the system. The significance is that this pressure mustn't be lower than the vapor pressure of the liquid being

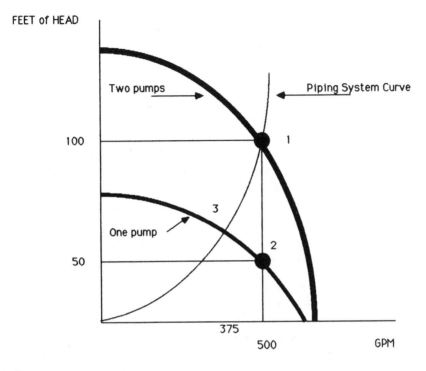

FEET of HEAD

Figure 12.7
SERIES PUMPS

POINT
1
The operating point on the system curve with both pumps on.
2
The actual point of operation of each pump wih both pumps on.
3
The operating point on the system curve if one pump is taken out of service.

pumped. Generally, if pumping is limited to air conditioning chilled water closed circuit systems there's usually no need being concerned with having enough pressure for good pump operation. It's also not ordinarily a factor in open systems or hot water systems unless there's considerable friction loss in the pipe or the water source is well below the pump and the suction lift is excessive. However, if the friction losses in the system are too great, meaning a lower than designed pressure at the pump suction, or the water temperature is high, the pressure at some point inside the pump may fall below the operating vapor pressure of the water, making the water boil and vaporize. This condition is called "cavitation" and it usually results in pitting and erosion of the impeller vane tips or inlet.

The problem of cavitation can be eliminated by maintaining a minimum suction pressure at the pump inlet to overcome the pump's internal losses. This minimum pressure is called net positive suction head. Net positive suction head is further broken down into required net positive suction head and available net positive suction head.

Required net positive suction head (NPSHR) refers to the internal pump losses and is a characteristic of the pump design. NPSHR for a specific pump is available from the manufacturer.

Available net positive suction head (NPSHA) is a characteristic of the system in which the pump operates. From a design standpoint, after the pressure loss in the system is determined, a safety factor is added to the NPSHR so that NPSHA exceeds NPSHR by 2 feet or more. From a field standpoint, if the pump is cavitating, or the measured NPSHA is insufficient, check the suction line for: undersized pipe, too many fittings, throttled valves, or clogged strainers.

Both available and required NPSH vary with capacity for a given pump and suction system. The NPSHA is decreased as the capacity is increased because of the increased friction losses in the suction piping. The NPSHR depends on the velocity and friction in the pump inlet and therefore, increases as the capacity increases.

PRIMARY-SECONDARY CIRCUITS

Primary-secondary circuits reduce pumping horsepower requirements while increaseing system control. When the two circuits are interconnected, the primary pump and the secondary pump will have no effect on each other; i.e., flow in one won't cause flow in the other, if the pressure drop in the common pipe is eliminated. The reason is the common piping, which may vary in length to a maximum of about 2 feet, has a negligible pressure drop and ensures the isolation of the secondary circuit from the primary circuit. Therefore, to overcome the pressure loss in the secondary circuit and provide design flow, a secondary pump must be installed. The primary pump circulates water around the primary circuit. The secondary pump supplies water to the terminals. The secondary flow may be less than, equal to, or greater than the primary flow (fig. 12.8).

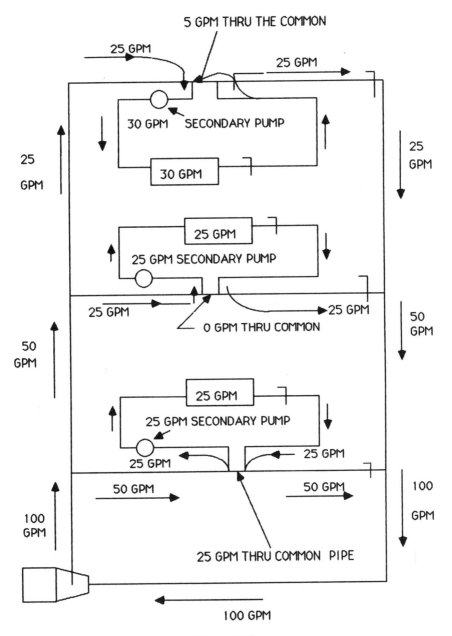

Figure 12.8

PART 3

Chapter 13
Fluid Flow, Psychrometric And Refrigeration Terminology

FLUID FLOW TERMINOLOGY

Acceleration Due to Gravity: The rate of increase in velocity of a body free falling in a vacuum. Equal to 32.2 ft/sec^2.

Fluid: A gas, vapor or liquid. Air is considered a compressible fluid and water a noncompressible fluid.

Fluid Dynamics: The condition of a fluid in motion. The velocity of a fluid is based on the cross-sectional area of the conduit and the volume of fluid passing through the conduit.

HVAC: Heating, Ventilating and Air Conditioning. Heating, ventilating and air conditioning a space using the fluids of air, water and refrigerants.

Nonuniform Flow: Fluid flow varying in velocity across the plane perpendicular to flow.

Turbulent Flow: Fluid flow in which the velocity varies in magnitude and direction in an irregular manner.

Uniform Flow: The smooth, straightline motion of a fluid across the area of flow.

PSYCHROMETRIC TERMINOLOGY

Adiabatic Process: A thermodynamic process when no heat is added to, or taken from, a substance.

Apparatus Dew Point (ADP): The temperature which would result if the psychrometric process line were carried to the saturation condition of the leaving air while maintaining the same sensible

171

heat ratio. The temperature of the coil at saturation. Also called effective coil temperature.

British Thermal Unit (Btu): The heat required to raise the temperature of one pound of water one degree Fahrenheit.

Density (d): Density is the weight of a substance per unit volume. Density is in pounds per cubic foot. Density is also the reciprocal of specific volume (d = 1/SpV).

Dew Point (DP): Dew point is the temperature at which moisture will start to condense from the air.

Dry Bulb Temperature (DB): Dry bulb temperature is the temperature of the air read on an ordinary thermometer. It's the measure of sensible temperature.

Enthalpy (h): Enthalpy is the measurement of the heat content of the air in Btu per pound of dry air.

Evaporative Cooling: The adiabatic exchange of heat between air and a water spray or wetted surface. The wet bulb temperature of the air remains constant but the dry bulb temperature is decreased.

Latent Heat (Ht): The heat which when supplied to or removed from a substance causes a change of state without any change in temperature. The units of latent heat are Btu per pound of dry air.

Relative Humidity (RH): Relative humidity is the ratio of the moisture present in the air to the total moisture that the air can hold at a given temperature. Relative humidity is expressed as a percentage.

Sensible Heat (Hs): The heat which causes a temperature change in a substance. The units of sensible heat are Btu per pound of dry air.

Sensible Heat Ratio (SHR): Sensible heat ratio is the ratio of the sensible heat to the total heat in the air (SHR = Hs/Ht).

Specific Heat (SpHt): The ratio of heat required to raise the temperature of one pound of substance 1 degree Fahrenheit as compared to the heat required to raise one pound of water 1 degree Fahrenheit. The specific heat of air is 0.24 Btu/lb/F. The specific heat of water is 1 Btu/lb/F.

Specific Humidity: Specific humidity, also called humidity ratio is the weight of water vapor associated with one pound of dry air. Specific humidity is measured in grains of moisture per pound of dry air or pounds of moisture per pound of dry air. 7000 grains equal one pound.

Specific Gravity (SpGr): The density of a substance compared to density of water. Density of water at 60 degrees Fahrenheit is 1.0.

Specific Volume (SpV): The volume of a substance per unit weight. Specific volume in cubic feet per pound. Specific volume is the reciprocal of density (SpV = 1/d).

Standard Air Conditions: Standard air is dry air having the following properties:
Temperature = 70 degrees F
Pressure = 29.92 inches mercury or 14.7 pounds per square inch.
Volume (specific volume) = 13.33 cubic feet per pound
Weight (specific density) = 0.075 pounds per cubic foot

Total Heat (Ht): Latent heat plus sensible heat. The units of total heat are Btu per pound of dry air.

Wet Bulb Depression: The temperature difference between the dry bulb temperature and the wet bulb temperature.

Wet Bulb Temperature (WB): Wet bulb temperature is obtained by an ordinary thermometer whose sensing bulb is covered with a wet wick and exposed to rapidly moving air. Wet bulb temperatures below 32 degrees Fahrenheit are obtained by an ordinary thermometer with a frozen wick.

REFRIGERATION TERMINOLOGY

Coefficient of Performance: Ratio of work performed to energy used.

Cooling Tower: A cooling tower cools heated water from a water-cooled condenser. As outdoor air passes through the cooling tower it removes heat from the condenser water. The water is cooled towards the wet bulb temperature of the entering air.

Compressor: The compressor is the pump in the refrigeration system. It takes the low temperature, low pressure refrigerant gas from the evaporator and compresses it to a high temperature, high pressure gas.

Condenser: The condenser receives the high temperature, high pressure refrigerant gas from the compressor and cools the gas to a high temperature, high pressure liquid. Condensers may be air-cooled or water-cooled.

Evaporator: The evaporator receives the high temperature, high pressure liquid from the condenser by way of the metering device. The metering device reduces the temperature and pressure of the liquid. In the evaporator, the low temperature, low pressure liquid is heated and changes state to a low temperature, low pressure gas. If the evaporator is in an air conditioning unit it's also called an evaporator coil, an evap coil, or a DX (dry-expansion/direct expansion) coil.

Filter-Drier: A combination device used as a strainer and moisture remover.

Metering Device: A capillary tube, automatic expansion valve (AEV, AXV), or thermostatic expansion valve (TEV, TXV). The metering device controls the flow of refrigerant and changes the high temperature, high pressure liquid refrigerant from the condenser to a low temperature, low pressure liquid.

Chapter 14
Air Systems:
Distribution, Components
And Terminology

Air Changes Per Hour: A method of expressing the quantity of air exchanged per hour in terms of conditioned space volume.

Air Conditioning: Treating or conditioning the temperature, humidity, and cleanliness of the air to meet the requirements of the conditioned space.

Air Conditioning Unit (ACU): An assembly of components for the treatment of air. Also called: air handling unit (AHU) for larger systems or fan-coil unit (FCU) for smaller systems.

Air Entrainment: The induced flow of the secondary or room air by the primary or supply outlet air, creating a mixed air path.

Airflow Patterns: Airflow patterns are important for proper mixing of supply and room air. Cooled air should be distributed from ceilings or high sidewall outlets. The airflow pattern for cooled air distributed from low sidewall or floor outlets should be adjusted to direct the air up. Heated air should be distributed from low sidewall or floor outlets. The airflow pattern for heated air distributed from ceilings or high sidewall outlets should be directed down.

Since most HVAC systems handle both heated and cooled air as the seasons change a compromise is made which usually favors the cooling season and ceiling or high sidewall outlets are specified. Generally, these outlets will be adjusted for a horizontal pattern. However, if the ceiling is very high or velocities are very low, such as when variable air volume boxes close off, it may be necessary to use a vertical pattern to force the air down to the occupied zone.

Ak Factor: The effective area of an air outlet or inlet.

Ambient Air: The surrounding air.

Aspect Ratio: Aspect ratio in rectangular ducts, air inlets or outlets is the ratio of width to height. Aspect ratio should not exceed 3:1.

Automatic Temperature Control(ATC) Dampers: Dampers controlled by temperature requirements of the system. ATC dampers should have a tight shutoff when closed. ATC dampers are usually opposed or parallel bladed dampers and can be either two-position or modulating. Two-position control means the damper is either open or closed. Example of two-position control: An automatic minimum outside air damper that's full open when the fan is on and shut when the fan is off.

   Modulating control provides for the gradual opening or closing of a damper. Example of modulating control: Automatic return and outside air dampers. The outside air damper closes as the return air damper opens and vice versa to allow more or less return or outside air to enter the mixed air plenum to satisfy the mixed air temperature requirements.

Backdraft Damper (BDD): A damper which opens when there's a drop in pressure across the damper in the direction of airflow and closes under the action of gravitational force when there's no airflow.

Balancing Station: An assembly to measure and control airflow. It has a measuring device and a volume control device with the recommended lengths of straight ductwork entering and leaving the station.

Ceiling Diffuser: A diffuser which typically provides a horizontal flow pattern that tends to flow along the ceiling producing a high degree of surface effect. Typical square or rectangular ceiling diffusers deliver air in a one, two, three or four-way pattern. Round ceiling diffusers deliver air in all directions.

Coil Bypass Factor (CBF): The coil bypass factor represents the amount of air passing through the coil without being affected by the coil temperature. It is a ratio of the leaving air dry bulb temperature minus the coil temperature divided by the entering air dry bulb temperature minus the coil temperature. A more efficient coil, that is a coil with more fins per inch and/or more rows of tubes, has a lower CBF.

Cold Deck: In a multizone or dual duct unit it's the chamber after the air leaves the cooling coil.

Comfort Zone: The range of effective temperatures and humidities over which the majority of adults feel comfortable. Generally, between 68F to 79F and 20% to 60% relative humidity.

Constant Volume Terminal Box: A terminal box which delivers a constant quantity of air. The boxes may be single duct or dual duct.

Constant Volume Single Duct Terminal Box: A single inlet terminal box supplied with air at a constant volume and temperature (typically cool air). Air flowing through the box is controlled by a manually operated damper or a mechanical constant volume regulator. The mechanical volume regulator uses springs and perforated plates or damper blades which decrease or increase the available flow area as the pressure at the inlet to the box increases or decreases. A reheat coil or cooling coil may be installed in the box or immediately downstream from it. A room thermostat controls the coil.

Constant Volume Dual Duct Terminal Box: A terminal box supplied by separate hot and cold ducts through two inlets. The boxes mix warm or cool air as needed to properly condition the space and maintain a constant volume of discharge air. Dual duct boxes may use a mechanical constant volume regulator with a single damper motor to control the supply air. The mixing damper is positioned by the motor in response to the room thermostat. As the box inlet pressure increases, the regulator closes down to maintain a constant flow rate through the box. Another type of constant flow regulation uses two motors, two mixing dampers and a pressure sensor to control flow and temperature of the supply air. The motor connected to the hot duct inlet responds to the room thermostat and opens or closes to maintain room temperature. The motor on the cold duct inlet is also connected to the room thermostat but through a relay which senses the pressure difference across the sensor. This motor opens or closes the damper on the cold duct inlet to (1) maintain room temperature and (2) maintain a constant pressure across the sensor and therefore, a constant volume through the box.

Damper: A device used to regulate airflow.

Deadband: A control scheme or controlled device that won't allow the mixing of hot and cold air.

Diffuser: A supply air outlet generally found in the ceiling with various deflectors arranged to promote mixing of primary air with secondary air. Types of diffusers are: round, square, rectangular, linear and troffers. Some diffusers have a fixed airflow pattern while others have field adjusted patterns.

Diversity: Variable Air Volume systems: The total volume (cubic feet per minute) of the VAV boxes is greater than the maximum output of the fan. For example, the VAV boxes total 40,000 cfm and the fan output is 30,000 cfm. Diversity is 0.75. Constant Air Volume systems: The total output of the fan is greater than the maximum required volume through the cooling coil. For example, the cooling coil is sized for 8,000 cfm and the fan has an output of 10,000 cfm. Diversity is 0.80.

Draft: A localized feeling of coolness caused by high air velocity, low ambient temperature, or direction of airflow.

Drop: The vertical distance that the lower edge of a horizontally projected air stream drops between the outlet and the end of its throw.

Dual Duct Terminal Box: A dual inlet terminal box supplied with any combination of heated, cooled, or dehumidified air. The hot duct supplies warm air which may be either heated air or return air from the conditioned space. The cold duct supplies cool air which may be either cooled and dehumidified when the refrigeration unit is operating or cool outside air brought in by the economizer cycle. A room thermostat controls a mixing damper arrangement in the boxes which determines whether the discharge air will be cool air, warm air or a mixture of both. Dual duct boxes are pressure independent and may be constant or variable volume.

Dual Path System: A system in which the air flows through heating and cooling coils essentially parallel to each other. The coils may be side-by-side or stacked. Multizone and dual duct systems are dual path. Some systems may not have a heating coil but instead bypass return air or mixed air into the hot deck.

Duct, Ductwork: A passageway made of sheet metal or other suitable material used for conveying air.

Dumping: The rapidly falling action of cold air caused by a variable air volume box or other device reducing airflow velocity.

Effective Area (Ak): The effective area of an outlet or inlet is the sum of the areas of all the vena contractas existing at the outlet and is affected by (1) the number of orifices and the exact location of the vena contractas and (2) the size and shape of the grille bars, diffuser rings, etc. Manufacturers have conducted airflow tests and based on their findings they've established flow factors or area correction factors for their products. Each flow factor, sometimes called "K-factor" or "Ak," applies to (1) a specific type and size of grille, register or diffuser, (2) a specific air measuring instrument and (3) the correct positioning of that instrument.

Equivalent Duct Diameter: The equivalent duct diameter for a rectangular duct.

Exhaust Air Inlet: An exhaust air grille, register or other opening to allow air from the conditioned space into the exhaust air duct.

Extractor: A device used in low pressure systems to divert air into branch ducts.

Face Velocity: The average velocity of the air leaving a coil, supply air outlet or entering a return or exhaust air inlet. For cooling coils, face velocity shouldn't exceed 600 fpm so condensate won't be blown off the coil. Recommended face velocities for outlets and inlets are given in the manufacturer's published data.

Fixed System: A fixed system is one in which there are no changes in the system resistance resulting from closing or opening of dampers, or changes in the conditions of filters or coils, etc. For a fixed system, an increase or decrease in system resistance results only from an increase or decrease in cfm and this change in resistance will fall along the system curve.

Grille: A wall-, ceiling- or floor-mounted louvered or perforated covering for an air opening. To control airflow pattern some grilles have a removable louver. Reversing or rotating the louver changes the air direction. Grilles are also available with adjustable horizontal or vertical bars so the direction, throw, and spread of the supply air stream can be controlled.

High Pressure Systems: Static pressures above 6 in. wg. with velocities above 2000 feet per minute.

Hot Deck: In a multizone or dual duct unit it's the chamber after the air leaves the heating coil.

Hunting: The condition which happens when a controller changes or cycles continuously resulting in fluctuation and loss of control. The desired set point condition can't be maintained.

Induction Terminal Box: Constant temperature supply air, called "primary air," is forced through a discharge device such as a nozzle or venturi to induce room or return air, "secondary air," into the box where it mixes with the supply air. The high velocity of the primary air creates a low pressure region which draws in or induces the higher pressure secondary air. The mixed air (primary + secondary) is then supplied to the conditioned space. Some induction boxes have heating or cooling coils through which the secondary air is induced. Induction boxes may be constant or variable volume, pressure dependent or pressure independent.

Light Troffers: A type of ceiling diffuser which fits over a fluorescent lamp fixture and delivers air through a slot along the edge of the fixture. They're available in several types. One type delivers air on both sides of the lamp fixture and another type provides air to only one side of the fixture.

Linear Slot Diffuser: This type of diffuser is manufactured in various lengths and numbers of slots and may be set for different throw patterns.

Low Pressure Systems: Static pressures to 2 in. wg. with velocities to 2500 feet per minute.

Make-Up Air: The air introduced to "make up" or replace air that's exhausted.

Medium Pressure Systems: Static pressures between 2 and 6 in. wg with velocities between 2000 and 4000 feet per minute.

Occupied Zone: The conditioned space from the floor to about 6 feet above the floor.

Opposed Blade Damper (OBD): A multiple bladed damper with a linkage which rotates the adjacent blades in opposite directions resulting in a series of openings that become increasingly narrow as the damper closes. This type of blade action results in a straight, uniform flow pattern sometimes called "non-diverting." Generally, opposed blade dampers are used in a volume control application.

Orifice Plate: An orifice plate is essentially a fixed circular opening in a duct. A measurable "permanent" pressure loss is created as the air passes from the larger diameter duct through the smaller opening. This abrupt change in velocity creates turbulence and a measurable amount of friction resulting in a pressure drop across the orifice. Calibration data which show flow rate in cubic feet per minute (cfm) versus measured pressure drop are furnished with the orifice plate. A differential pressure gage such as a manometer is connected to the pressure taps and flow is read.

Outlet Velocity: The average velocity of air emerging from a fan, outlet or opening.

Parallel Blade Damper: Generally, parallel blade dampers are used in a mixing application. Because the blades rotate parallel to each other, a parallel blade damper produces a "diverting" type of air pattern and when in a partially closed position, the damper blades throw the air to the side, top or bottom of the duct. This flow pattern may adversely affect coil or fan performance or the air-flow into branch ducts if the damper is located too close upstream.

Perforated Face Diffuser: Perforated face diffusers are used with lay-in ceilings and are similar in construction to the standard square ceiling diffuser with an added perforated face plate. They're generally equipped with adjustable vanes to change the flow pattern to a one-, two-, three-, or four-way throw.

Plenum: An air chamber or compartment.

Pressure Dependent Terminal Box: The quantity of air passing through the box is dependent on the inlet static pressure.

Pressure Independent Terminal Box: The quantity of air passing through the box is independent (within design limits) of the inlet static pressure.

Radius of Diffusion: The horizontal distance an air stream travels after leaving the outlet before it's reduced to its specified terminal velocity.

Register: A grille with a built-in or attached damper assembly.

Residual Velocity: Room velocity.

Room Velocity: The air velocity in the occupied zone.

Return Air Inlet: A return air grille, register, or other inlet such as a perforated face opening, linear slot, light troffer, or other opening to allow air from the conditioned space or return air plenum into a return air duct or mixed air plenum. Inlets are generally chosen and located to suit architectural design requirements for appearance and compatibility with supply outlets. In most cases, the location of inlets doesn't significantly affect air motion and temperature except when the inlet is positioned directly in the primary air stream from the outlet. This "short circuits" the supply air back into the return system without properly mixing with the room air.

Single Duct Terminal Box: A terminal box usually supplied with cool air through a single inlet duct. The boxes may be constant or variable volume, pressure dependent or pressure independent. They may also contain a water coil (heating or cooling), steam coil, or electric reheat.

Single Path: A system in which the air flows through coils essentially in series to each other. Single zone heating and cooling units and terminal reheat units are examples.

Smudging: The black markings on ceilings and outlets usually made by suspended dirt particles in the room air which is then entrained in the mixed air stream and deposited on the ceilings and outlets. Anti-smudge rings are available which lower the outlet away from the ceiling and cover the ceiling area a few inches beyond the diffuser.

Splitter Damper: A device used in low pressure systems to divert airflow. Splitter damper is a misnomer. It's not a damper for regulating airflow. It's a diverter for directing the airflow.

Spread: The divergence of the air stream after it leaves the outlet.

Stratification: Layers of air at different temperatures or different velocities flowing through a duct or plenum. Also called stratified air.

Supply Air Outlet: A supply air diffuser, grille, register, or other opening to allow supply air into the conditioned space to mix with the room air to maintain a uniform temperature throughout the occupied zone. Supply air diffusers, grilles, and registers, are chosen and located to control airflow patterns to avoid drafts and air stagnation, and to complement the architectural design of the building.

Surface Effect: The effect caused by entrainment of secondary air when an outlet discharges air directly parallel and against a wall or ceiling. Surface effect is good for cooling applications, especially variable air volume systems because it helps to reduce the dumping of cold air. Surface effect contributes to smudging.

Terminal Box: A device or unit which regulates supply airflow, temperature and humidity to the conditioned space. Terminal boxes are classified as single duct, dual duct, constant volume, variable volume, medium pressure, high pressure, pressure dependent, pressure independent, system powered, fan powered, induction, terminal reheat and bypass. They may also contain a combination of heating or cooling coils, dampers and sound attenuation. The airflow through the box is normally set at the factory but can also be adjusted in the field. Terminal boxes also reduce the inlet pressures to a level consistent with the low pressure, low velocity duct connected to the discharge of the box. Any noise that's generated within the box in the reduction of the pressure is attenuated. Baffles or other devices are installed which reflect the sound back into the box where it can be absorbed by the box lining. Commonly, the boxes are lined with fiber glass which also provides thermal insulation so the conditioned air within the box won't be heated or cooled by the air in the spaces surrounding the box. Terminal boxes work off static pressure in the duct system. Each box has a minimum inlet static pressure requirement to overcome the pressure losses through the box plus any losses through the discharge duct, volume dampers, and outlets.

Terminal Velocity: The maximum air velocity of the mixed air stream at the end of the throw.

Throw: The horizontal and vertical distance an air stream travels after leaving the outlet before it's reduced to its specified terminal velocity.

Variable Air Volume (VAV) Terminal Box: VAV boxes are available in many combinations that include: pressure dependent, pressure independent, single duct, dual duct, cooling only, cooling with reheat, induction, bypass and fan powered. VAV boxes can also be classified by (1) volume control: throttling, bypass, or fan powered, (2) intake controls and sensors: pneumatic, electric, electronic, or system powered, (3) thermostat action: direct acting or reverse acting and (4) the condition of the box at rest: normally open or normally closed. The basic VAV box has a single inlet duct. The quantity of air through the box is controlled by throttling an internal damper. If the box is pressure dependent, the damper will be controlled just by a room thermostat, whereas, the pressure independent version will also have a regulator to limit the air volume between a preset maximum and minimum.

Inside the pressure independent box is a sensor. Mounted on the outside is a controller with connections to the sensor, volume damper and room thermostat. The quantity of air will vary from a design maximum cfm down to a minimum cfm which is generally around 50% to 25% of maximum, but may be as low as zero (shutoff). The main feature of the VAV box is its ability to vary the air delivered to the conditioned space as the heating load varies. Then, as the total required volume of air is reduced throughout the system, the supply fan will reduce its cfm output. This means a savings of energy and cost to operate the fan. The exception to this is the VAV bypass box.

The types of controls used to regulate the flow of air through VAV boxes are as varied as the types of boxes. Many boxes are designed to use external sources of power: pneumatic, electric or electronic. These boxes are sometimes called non-system powered. Other boxes are system powered which means that the operating controls are powered by the static pressure from the main duct system. System powered boxes don't need a separate pneumatic or electric control system. This reduces first costs, however, they usually have a higher required minimum inlet static pressure which means that the supply fan will be required to produce higher static pressures, resulting in increased operating costs. All controllers, except for the bypass box type, reduce airflow.

VAV Bypass Terminal Box: A bypass box uses a constant volume supply fan but provides variable air volume to the conditioned space. The supply air comes into the box and can exit either into the conditioned space through the discharge ductwork or back to the return system through a bypass damper. The conditioned space receives either all the supply air or only a part of it depending on what the room thermostat is calling for. Since there's no reduction in the main supply air volume feeding the box, this type of system has no savings of fan energy.

VAV Ceiling Induction Terminal Box: A ceiling VAV induction box has a primary damper at the box inlet and an induction damper in the box which allows air in from the ceiling plenum. On a call for cooling, the primary damper is full open and the induction damper is closed. As the conditioned space cools down the primary damper throttles back and the induction damper opens to maintain a constant mixed airflow to the conditoned space. At some point, the induction damper is wide open and the primary damper is throttled (about 75%) to allow for the maximum induction ratio. Another type of induction box has a constant pressure nozzle inducing either primary air from the main supply system or return air from the ceiling plenum. The room thermostat opens or closes a primary air bypass damper to allow for the induction of primary or return air. This box uses volume regulators to reduce the airflow to the conditioned space. Some boxes may contain a reheat coil.

VAV Dual Duct Terminal Box: A VAV dual duct terminal box is supplied by separate hot and cold ducts through two inlets. A variety of control schemes vary the air volume and discharge air temperature. One type uses a temperature deadband which supplies a varying quantity of either warm or cool air, but not mixed air, to the conditioned space.

VAV Fan Powered Terminal Box: A VAV fan powered box has the advantage of the energy savings of a conventional, single duct VAV system with the addition of several methods of heating and a constant airflow to the conditioned space. The box contains a fan and a return air opening from the ceiling space. When the room thermostat is calling for cooling the box operates as would the standard VAV box. However, on a call for heat the fan draws warm (secondary) air from the ceiling plenum and recirculates it into the rooms. Varying amounts of cool (primary) air from the main system are introduced into the box on either the inlet or discharge side of the fan and mix with the secondary air. A system of dampers, backdraft or motorized, control the airflow and mixing of the air streams. As the room thermostat continues to call for heat, the primary air damper closes off and more secondary air is drawn into the box and it alone is recirculated. Therefore, the airflow to the conditioned space stays constant. If more heat is needed, reheat coils may be installed in the boxes. The fan may operate continuously or it may shut off. A common application of fan powered boxes is around the perimeter or other areas of a building where: (1) air stagnation is a problem when the primary air throttles back, (2) zones have seasonal heating and cooling requirements, (3) heat is needed during unoccupied hours when the primary fan is off, (4) heating loads can be offset mainly with recirculated return air.

VAV Fan Powered Bypass Terminal Box: This box acts the same as the conventional bypass box with the addition of a secondary fan in the box. The bypass box uses a constant volume supply primary fan but provides variable air volume to the conditioned space. The supply air comes into the box and can exit either into the conditioned space through the secondary fan and the discharge ductwork or back to the return system through a bypass damper. The fan in the box circulates the primary air or return air into the room. The conditioned space receives either all primary air, all return air or a mixture of the two, depending on what the room thermostat is calling for. Since there's no reduction in the main supply air volume feeding the box, this type of system has no savings of primary fan energy.

VAV Pressure Dependent Terminal Box: A pressure dependent VAV box is essentially a pressure reducing and sound attenuation box with a motorized damper that's controlled by a room thermostat. These boxes don't regulate the airflow, but simply position the damper in response to the signal from the thermostat. Because the airflow to these boxes is in direct relation to the box inlet static pressure, it's possible for the boxes closest to the supply fan, where the static pressure is the greatest, to get more air than is needed, so the boxes farther down the line will be getting little or no air. Therefore, pressure dependent boxes should only be installed in systems where there's no need for limit control and the system static pressure is stable enough not to require pressure independence.

Pressure dependent maximum regulated volume boxes may be used where pressure independence is required only at maximum volume and the system static pressure variations are only minor. These boxes regulate the maximum volume but the flow rate at any point below maximum varies with the inlet static pressure. This may cause "hunting."

VAV Pressure Independent Terminal Box: Pressure independent VAV boxes can maintain airflow at any point between maximum and minimum, regardless of box inlet static pressure, as long as the pressure is within the design operating range. Flow sensing devices regulate the flow rate through the box in response to the room thermostat's call for cooling or heating.

VAV Single Duct Pressure Independent Terminal Box: To maintain the correct airflow in a pressure independent box over the entire potential range of varying inlet static pressure, a sensor reads the differential pressure at the inlet of the box and transmits it to the controller. The room thermostat responding to the load conditions in the space also sends a signal to the controller. The controller responds by actuating the volume damper and regulating the airflow within the preset maximum and minimum range. For example, as the temperature rises in the space, the damper opens for more cooling. As the temperature in the space drops, the damper closes. If the box also has a reheat coil, the volume damper, on a call for heating, would close to its minimum position (usually not less than 50% of maximum) and the reheat coil would be activated. Because of its pressure independence, the airflow through the boxes is unaffected as other VAV boxes in the system modulate and change the inlet pressures throughout the system.

VAV System Powered Terminal Box: System powered VAV boxes use the static pressure from the supply duct to power the VAV controls. The minimum inlet static pressure with this type of box is usually higher than other VAV systems in order to (1) operate the controls and (2) provide the proper airflow quantity.

Vena Contracta: The smallest area of an air stream leaving an orifice.

Ventilation: Supplying air to or removing air from a space by natural or mechanical means.

Venturi: A Venturi operates on the same principle as the orifice plate but its shape allows gradual changes in velocity and the "permanent" pressure loss is less than is created by an orifice plate. Calibration data which show flow rate in cubic feet per minute (cfm) versus measured pressure drop are furnished with the Venturi. The pressure drop is measured with a differential gage.

Volume Dampers (VD): Manual dampers used to control the quantity of airflow in the system by introducing a resistance to flow. If not properly selected, located, installed and adjusted, they (1) don't control the air as intended, (2) they add unnecessary resistance to the system and (3) they can create noise problems. The resistance a volume damper creates in a duct system is determined by how complicated the system is. For instance, if the system is very simple and the damper is a large part of that resistance, then any movement of the damper will change the resistance of the entire system and good control of the airflow will result. If, however, the damper resistance is very small in relation to the entire system, poor control will be the case. For instance, partial closing of a damper will increase its resistance to airflow, but depending on the resistance of the damper to the overall system resistance, the reduction in airflow may or may not be in proportion to closure. In other words, closing a damper 50% doesn't necessarily mean that the airflow will be reduced to

50%. For example, a damper when open might be 10% of the total system resistance. When this damper is half closed the airflow will be reduced to 80% of maximum flow. However, a similarly built damper in another duct system is 30% of the total system resistance when open. When this damper is half closed the airflow is reduced to 55% of maximum.

The relationship between the position of a damper and its percent of airflow is termed its "flow characteristic." Opposed blade dampers are generally recommended for large duct systems because they introduce more resistance to airflow for most closed positions, and therefore, have a better flow characteristic than parallel blade dampers. However, flow characteristics of dampers aren't consistent and may vary from one system to another. The actual effect of closing a damper can only be determined in the field by measurement.

Proper location of balancing dampers not only permits maximum air distribution but also equalizes the pressure drops in the different airflow paths within the system. Manual dampers should be provided in each takeoff to the runout to control the air to grilles and diffusers. They should also be in (1) the main, (2) each submain, (3) each branch and (4) each subbranch duct. Manually operated opposed blade or single blade quadrant type volume dampers should also be installed in every zone duct of a multizone system.

Single blade or opposed blade volume dampers immediately behind diffusers and grilles shouldn't be used for balancing because when throttled they (1) create noise at the outlet and (2) change the effective area of the outlet so the flow (Ak) factor is no longer valid. Proper installation and location of balancing dampers in the takeoffs eliminate the need for volume controls at grilles and diffusers.

Manual volume dampers may need to be installed in the outside, relief and return air connections to the mixed air plenum in addition to any automatic dampers. These volume control dampers balance the pressure drops in the various flow paths so the pressure drop in the entire system stays constant as the proportions of return air and outside air vary to satisfy the temperature requirements.

Volume dampers and handles should have enough strength

and regidity for the operating pressures of the duct system in which they will be installed. For small duct, a single blade damper is satisfactory. For large duct, dampers should be opposed blade. Each damper should have a locking quadrant handle or regulator.

Chapter 15
Water Systems:
Distribution, Components
And Terminology

Air Separators: Air separators free the air entrained in the water in a hydronic system. There are several types of air separators. The centrifugal type of air separator works on the action of centrifugal force and low velocity separation. The centrifugal motion of the water circulating through the air separator creates a vortex or whirlpool in the center of the tank and sends heavier, air-free water to the outer part of the tank and allows the lighter air-water mixture to move to a low velocity air separation and collecting screen located in the vortex. The entrained air, being lighter, collects and rises into the compression tank. The boiler dip tube type of air separator is a tube in the top or top side of the boiler. When the water is heated, air is released and collects at a high point in the boiler. The dip tube allows this collected air to rise into the compression tank. The in-line, low velocity air separating tank with dip tube type of air separator is used when a boiler isn't available or not useable as the point of air separation.

Annular Flow Meters: Annular flow meters have a multi-ported flow sensor installed in the water pipe. The holes in the sensor are spaced to represent equal annular areas of the pipe in the same manner as the Pitot tube traverse is for round duct. The flow meter is designed to sense the velocity of the water as it passes the sensor. The upstream ports sense high pressure and the down-stream ports sense low pressure. The resulting difference, or differential pressure is then measured with an appropriate differential pressure gage. Calibration data which show flow rate in gallons per minute (gpm) versus measured pressure drop is furnished with the flow meter.

Automatic Air Vents: There are two types of automatic air vents. The hydroscopic type contains material that expands when wet and holds the air vent valve closed. Air in the system dries out the material, causing it to shrink and opening the vent valve. The float type contains a float valve that keeps the air vent closed while there's water in the system. If air is in the system, it rises to the air vent. When it reaches the float, the float drops and opens the air vent valve.

Automatic Control Valves (ACV): Automatic control valves are used to control flow rate or to mix or divert water streams. They're classified as two-way or three-way construction and either modulating or two-position.

Balancing Station: An assembly to measure and control water flow. It has a measuring device and a volume control device with the recommended lengths of straight pipe entering and leaving the station.

Ball Valve: The ball valve is a manual valve used for regulating water flow. It is similar to the plug valve and is often used for water balancing. It has a low pressure drop and good flow characteristics.

Butterfly Valve: The butterfly valve is a manual valve used for regulating water flow. It has a heavy ring enclosing a disc which rotates on an axis and in principle, is similar to a round single blade damper. It has a low pressure drop and is used as a balancing valve but it doesn't have the good throttling characteristics of a ball or plug valve.

Calibrated Balancing Valves: With the other types of flow meters such as the orifice plate, Venturi and annular type, a balancing valve is also needed to set the flow. Calibrated balancing valves are designed to do both duties of a flow meter and a balancing valve. These valves are similar to ordinary balancing valves except the manufacturer has provided pressure taps in the inlet and outlet and has calibrated the device by measuring the resistance at various valve positions against known flow quantities. The valve has a graduated scale or dial to show the degree open. Calibration data which show flow rate in gallons per minute (gpm) versus measured pressure drop are provided with the valve. Pressure drop is measured with any appropriate differential gage.

Check Valve: The check valve is a manual valve which limits the direction of flow. The check valve allows water to flow in one direction and stops its flow in the other direction. The swing check valve has a gate which opens when the system is turned on and pressure from the water flowing in the proper direction is applied. The gate closes due to its own weight and gravity when the system is off. The spring loaded check valve has a spring that keeps the valve closed. Water pressure from the proper direction against the spring opens the valve.

Closed System: A closed system is when there's no break in the piping circuit and the water is closed to the atmosphere.

Coils: Coils are heat transfer devices (heat exchangers). They come in a variety of type and sizes and are designed for various fluid combinations. In hydronic applications coils are used for heating, cooling or dehumidifying air. Hydronic coils are most often made of copper headers and tubes with aluminum or copper fins and galvanized steel frames.

Coil Face Area: The area (width x height) across which the air flows.

Cooling Coil: A chilled water or refrigerant coil.

Combination Valve: The combination valve regulates flow and limits direction. It's a combination of a check valve, calibrated balancing valve and shutoff valve. It's made in a straight or angle pattern. The valve acts as a check valve preventing backflow when the pump is off and can be closed for tight shutoff or partly closed for balancing. The valve has pressure taps for connecting flow gages and reading pressure drop across it. A calibration chart is supplied with the valve for conversion of pressure drop to gpm. The valve also has a memory stop. Combination valves are sometimes called multi-purpose or triple-duty valves.

Counter Flow Coils: Coils are piped counter flow for the greatest heat transfer for a given set of conditions. Counter flow means that the flow or air and water are in opposite directions to each other. The water enters on the same side that the air leaves. For cooling coils, this would mean that the coldest water is entering the coil at the point that the coldest air is leaving the coil. This allows the greatest heat transfer.

Dual Temperature Water (DTW): A typical chilled water range is 45-55 degrees. A typical heating water temperature range is

100-150 degrees.

Evaporator Coil: A coil in the air handling unit containing a refrigerant such as R-22 or R-12. Also called a refrigerant coil or simply, an evaporator.

Expansion/Compression Tanks: A tank which compensates for the normal expansion and contraction of water in a hydronic system. Water expands when heated in direct proportion to its change in temperature. In hydronic systems, an allowance must be made for this expansion, otherwise when the system is completely filled, the water has nowhere to go and there's the possibility of a pipe or piece of equipment breaking.

In an open system, an expansion tank which is open to the atmosphere, is provided about 3 feet above the highest point in the system. Then as the water temperature increases (and therefore, the total volume of water in the system increases) the water level rises in the tank. Open expansion tanks are limited to installations having operating temperatures of 180 degrees F or less because of water boiling and evaporation problems. Being open to the atmosphere also has the drawback of the continual exposure to the air and its possible corrosive effects.

In a closed system a closed expansion tank containing air is used. Now when the water expands it partially fills the tank compressing the air. Closed expansion tanks are also known as compression tanks. Pressure in the system will vary from the minimum pressure required to the maximum allowable working pressure. The minimum pressure requirement, anywhere in the system, must always be (1) greater than the vapor pressure of the water to avoid cavitation problems and (2) greater than atmospheric so air doesn't enter the system. Maximum allowable working pressure (MAWP) must not exceed the construction limits. On a low temperature boiler, for example, the pressure relief valve is set for 30 psig which is therefore, the maximum allowable working pressure.

When the system is filled with water the air in the compression tank (about 2/3 of the tank is water, 1/3 is air) will act as a spring or cushion to keep the proper pressure on the system to accommodate the fluctuations in water volume and control pressure change in the system. If there's inadequate air control in the system, the tank might become "waterlogged." This happens when the water in the system is heated. The air in the heated water is released and travels up the system and is vented out at the high points. When the boiler is off the water in the compression tank cools. The cool water absorbs air from the tank and flows into the boiler by gravity. When the water is heated again the air is again released and vented out. After several cycles, all the air is removed from the tank and the tank fills with water. The water now has nowhere to go except the pressure relief valve on the boiler which will begin opening on every boiler firing cycle to relieve the pressure caused by the water expansion.

If the air leaks out of the compression tank (check for leaks around the sight glass), the pressure on the system is reduced below the setpoint on the pressure reducing valve (PRV). The pressure reducing valve will open to allow in more water to fill the system and the tank until the setpoint on the PRV is reached. When the water in the system is heated it expands to fill the tank since there's no air cushion. This results again in a waterlogged tank. The pressure relief valve will spill water every time the boiler fires. When the pressure relief valve opens it reduces the pressure in the system and again the PRV opens to bring fresh water into the system.

Compression tanks should be installed on the suction side of the pump. The point where the compression tank connects to the system is called "the point of no pressure change."

Fixed System: A fixed system is one in which there are no changes in the system resistance resulting from closing or opening of valves, or changes in the condition of the coils, etc. For a fixed system, an increase or decrease in system resistance results only from an increase or decrease in gpm and this change in resistance will fall along the system curve.

Flexible Connectors: Flexible connectors are used between piping and pumps and other pieces of equipment to reduce noise and vibration. Flexible connectors can be made of rubber, plastic or braided metal.

Four-Pipe Main: The four-pipe arrangement is used where independent heating and cooling is called for. The four-pipe system is two separate two-pipe arrangements. One two-pipe arrangement for chilled water and one for heating water. No mixing occurs. The return connections from the terminals can be made either direct or reverse return.

The air handling unit is usually provided with two separate water coils, one for heating and one for chilled water. Each coil has its own control valve. However, some air handling units have only one coil. When this is the case, modulating three-way valves, like the ones used in the three-pipe system, are installed in the supply branches. A three-way, two-position valve on the return branch line diverts the leaving hot or cold water to the proper main.

Flow Meter: Flow meters such as orifice plates, Venturis, annular type and calibrated balancing valves are permanently installed devices used for flow measurements of pumps, primary heat exchange equipment, distribution pipes and terminals. For flow meters to give accurate, reliable readings they should be installed far enough away from any source of flow disturbance to allow the turbulence to subside and the water flow to regain uniformity. The manufacturers of flow meters usually specify the lengths of straight pipe upstream and downstream of the meter needed to get good readings. Straight pipe lengths vary with the type and size of flow meter but typical specifications might be between 5 to 25 pipe diameters upstream and 2 to 5 pipe diameters downstream of the flow meter.

Fins: Fins on a coil increase the area of heat transfer surface to improve the efficiency and rate of transfer and are generally spaced from 4 to 14 fins per inch (fpi). As with coil tubes, the more fins, the more heat transfer, but also the more resistance to airflow. Aluminum is usually picked over copper for fin material for reasons of economy. However, when cooling coils are sprayed with water, cooper fins are needed to prevent electrolysis between the dissimilar copper tubes and aluminum fins. Coils wetted only by condensation are seldom affected by electrolysis and are usually copper tube, aluminum fin.

Gate Valve: The gate valve is a manual valve used for tight shutoff to service or remove equipment. It has a straight through flow passage which results in a low pressure drop. It regulates flow only to the extent that it's either fully open or fully closed.

Don't use it for throttling purposes because the internal construction is such that when it's partly closed the resulting high velocity water stream will cause erosion of the valve seat. This erosion or "wiredrawing" will lead to eventual leakage when the valve is fully closed.

Globe Valve: The globe valve is a manual valve used in water make-up lines. It can be used in partially open positions and therefore, can be used for throttling flow. However, the globe valve has a high pressure drop even when fully open which unnecessarily increases the pump head and therefore, shouldn't be used for balancing.

Heat Exchanger: A device specifically designed to transfer heat between two physically separated fluids. The term heat exchanger can describe any heat transfer device such as a coil or a particular category of devices. Heat exchangers are made in various sizes and types (shell and tube, U-tube, helical and plate) and are designed for several fluid combinations such as steam to water (converter), water to steam (generator), refrigerant to water (condenser), water to refrigerant (chiller), water to water (heat exchanger), air to refrigerant (coil), and air to water or water to air (coil).

Heating Coil: A hot water coil.

High Temperature Water (HTW): Temperature range of 350 to 450 degrees.

Hydronics: The science of heating and cooling with liquids.

Log Mean Temperature Difference (LMTD): The logarithmic average of the temperature differences when two fluids are used in a heat transfer process. The temperature difference after the process will be less than at the beginning and the exchange of heat will follow a logarithmic curve. It's this logarithmic average of the temperature differences which establishes which heat exchanger is best. The higher the LMTD number the greater the heat transfer. For example, a coil piped parallel flow has a lower LMTD and would need more surface area to do the same heat transfer as a coil piped counter flow which has a higher LMTD.

Low Temperature Water (LTW): Temperature range to 250 degrees.

Make-Up Water: The water that replaces the water lost through leakage and evaporation. To prevent air problems the make-up water to a closed system should be introduced into the system at some point either in the air line to the compression tank or at the bottom of the compression tank.

Manual Air Vent: A manual valve which is opened to allow entrained air to escape.

Manual Valves: Manual valves are used to regulate flow rate or limit the direction of flow. Gate, globe, plug, ball and butterfly valves regulate flow rate while a check valve limits the direction of flow. A valve called a combination valve does both.

Medium Temperature Water (MTW): Temperature range of 250 to 350 degrees.

Memory Stops: Some plug valves, calibrated balancing valves and combination valves have adjustable memory stops. The memory stop is set during the final balance so that, if the valve is closed for any reason, it can later be reopened to the original setting.

One-Pipe Main: The one-pipe main is used for individual space control in residential and small commercial and industrial heating applications. This piping arrangement uses a single loop main but differs from the series loop arrangement since each terminal is connected by a supply and return branch pipe to the main. Because the terminal has a higher pressure drop than the main, the water circulating in the main will tend to flow through the straight run of the tee fittings. This starves the terminal. To overcome this problem a diverting tee is installed in either the supply branch, return branch or sometimes, both branches. Diverting tees create the proper resistance in the main to direct water to the terminal. The advantage of the one-pipe main arrangement over the series loop is that each terminal can be separately controlled and serviced by installing the proper valves in the branches. However, as with the series loop arrangement, if there are too many terminals the water temperature at the terminals farthest from the boiler may not be adequate. To have the water temperature to each terminal equal to the temperature at which it's generated, two-pipe arrangements are used.

Open System: An open system is when there's a break in the piping circuit and the water is open to the atmosphere.

Orifice Plate: A hydronic orifice plate is essentially a fixed circular opening in a pipe. A measurable "permanent" pressure loss is created as the water passes from the larger diameter pipe through the smaller opening. This results in an abrupt change in velocity causing turbulence and a measurable amount of friction which results in a pressure drop across the orifice. Calibration data which show flow rate in gallons per minute (gpm) versus measured pressure drop are furnished with the orifice plate. A water gage such as a differential pressure gage is connected to the pressure taps and flow is read.

Parallel Flow Coils: Parallel flow means that the flow of air and water are in the same direction to each other. The water and air enter on the same side. For parallel flow cooling coils, the coldest water enters the coil at the point where the warmest air enters the coil, therefore, less heat transfer.

Pete's Plug: Pete's Plugs® are test points installed in the piping on the entering and leaving sides of chillers, condensers, boilers and coils to take temperatures or pressures. A hand held thermometer or pressure probe is inserted.

Plug Valve: Plug valves are manual valves used for balancing water flow. Plug valves have a low pressure drop and good throttling characteristics and therefore, add little to the pumping head. Plug valves can also be used for tight shutoff.

Pressure Reducing Valves (PRV): The PRV reduces the pressure from the city water main to the proper pressure needed to (1) completely fill the system and (2) maintain this pressure. Pressure reducing valves are installed in the piping that supplies make-up water to the system. They generally come set at 12 psi (about 28 feet). This is adequate for one- and two-story buildings; however, for three-story or higher buildings, the PRV should be adjusted so there's a minimum of 5 psi additional pressure at the highest terminal. For example, if a coil were at an elevation of 30 feet, the PRV should be set for 18 psi (30' is equal to 13 psi + 5 psi safety). To change the adjustment of the PRV, remove the cap from the top of the valve, loosen the jam nut, and turn the adjusting screw in a clockwise direction. This increases the pressure on the system. Turning the adjusting screw in a counter-clockwise direction lowers the pressure.

Pressure Relief Valves: Pressure relief valves are safety devices to protect the system and human life. The pressure relief valve opens on a preset value so the system pressure can't exceed this amount.

Primary-Secondary Circuits: Primary-secondary circuits reduce pumping horsepower requirements while increasing system control. The primary pump and the secondary pump have no effect on each other when the two circuits are properly interconnected. Flow in one circuit won't cause flow in the other if the pressure drop in the pipe common to both circuits is eliminated. The secondary flow may be less than, equal to, or greater than the primary flow.

Series Loop: The series loop piping arrangement is generally limited to residential and small commercial heating applications. In a series loop, supply water is pumped through each terminal in series and then returned back to the boiler. The advantages to this type of piping arrangement are it's simple and inexpensive. The disadvantages are: (1) if repairs are needed on any terminal the whole system must be shut down and (2) it's not possible to provide a separate capacity control to any individual terminal since valving down one terminal reduces flow to the terminals down the line. However, space heating can be controlled through dampering airflow. These disadvantages can be partly remedied by designing the piping with two or more circuits and installing balancing valves in each circuit. This type of arrangement is called a split series loop.

The series loop circuit length and pipe size are also important because they directly influence the water flow rate, temperature and pressure drop. For instance, as the heating supply water flows through the terminals, its temperature drops continuously as it releases heat in each terminal. If there are too many terminals in series, the water temperature in the last terminal may be too cool.

Strainer: Strainers catch sediment or other foreign material in the water. A strainer contains a fine mesh screen formed into a sleeve or basket that fits inside the strainer body. This sleeve must be removed and cleaned. A strainer with a dirty sleeve or a sleeve with a screen that's too fine means there will be excessive pressure drop across the strainer and lower water flow. Individual fine mesh sleeve strainers may also be installed before automatic control valves or spray nozzles which operate with small clearances and need protection from materials which might pass through the pump strainer.

System Designations: Some system designations are: Heating Water (HW), Heating Water Supply (HWS), Heating Water Return (HWR), Chilled Water (CHW), Chiller Water Supply (CHWS), Chilled Water Return (CHWR), Condenser Water (CW), Condenser Water Supply (CWS), Condenser Water Return (CWR).

Temperature Wells: Temperature wells are installed at specific points in the piping so a test thermometer can be inserted to measure the temperature of the water in the pipe. Generally, temperature wells are installed on the entering and leaving sides of chillers, condensers, boilers and coils. Thermometer wells must be long enough to extend into the pipe so good contact is made with the water. The well forms a cup to hold a heat-conducting liquid (usually an oil) so good heat transfer is made from the water in the pipe to the liquid in the well to the thermometer. Therefore, wells should be installed vertically or not more than 45 degrees from vertical, so they'll hold the liquid. You will not get an accurate reading from inserting a thermometer into a dry well, as air will act as an insulator.

Three-Pipe Main: A three-pipe arrangement has two supply mains and one return main.

One supply circulates chilled water from the chiller and the other supply circulates heating water from the boiler. This permits any space to be independently cooled or heated. A three-way valve in each supply branch switches to deliver either chilled water or heating water, but not both, to the terminal. The supply flows aren't mixed.

The return main, however, receives water from each terminal. This means that frequently the return will be handling a mixture of chilled and hot waters. This results in a waste of energy as both the chiller and the boiler receive warm water and must work harder to supply their proper discharge temperatures. The return connections from the terminals can be made either direct or reverse return.

Three-Way Automatic Control Valves: Three-way ACV's are used to mix or divert water flow and are generally classified as mixing or diverting valves. They may be either single seated (mixing valve) or double seated (diverting valve). The single seated, mixing valve is the most common. The terms "mixing" and "diverting" refer to the *internal construction of the valve and not the application.* The internal difference is necessary so the valve will seat against flow (see two-way valves). A mixing valve has two inlets and one outlet. The diverting valve has one inlet and two outlets. Either valve may be installed for a flow control action (bypassing application) or a temperature control action (mixing application) depending on its location in the system. Diverting valves, however, shouldn't be substituted for mixing valves and vice versa. Using either design for the wrong application would tend to cause chatter.

Another type of modulating three-way valve is used in the supply line to coils in a three-pipe system. This valve has two inlets and one outlet. One inlet is supplied with heating water and the other is supplied with chilled water. The valve varies the quantity of heating and chilled water but doesn't mix the two streams. Depending on the thermostatic controls in the occupied space, the valve opens to allow either heating water only or chilled water only into the coil. This same type of three-way modulating valve is used in the supply line to a four-pipe, one-coil system. The return line also has a three-way valve, but it's two position; not modulating. The return valve has one inlet and two outlets. The water leaving the coil enters the valve and is diverted to either the heating water return main or the chilled water return main depending on temperature of the water entering the coil. For example, if the supply valve is allowing heating water into the coil, the return valve will be diverting this water to the heating return main.

Throttling Characteristics: Throttling characteristics refer to the relationship of the position of the valve disc and its percent of flow. A valve has a linear throttling characteristic when the disc open percentage is the same as the flow percentage. An example of a valve with a linear or straight line throttling characteristic would be a disc that's 50% open and the flow is measured at 50%. Then if the disc is closed to 30% open, the flow would be reduced to 30%.

Tubes: Coil tubes are usually made of copper but other materials used include carbon steel, stainless steel, brass and for special applications, cupro-nickel. For applications where the air stream may contain corrosives, there are various protective coatings available. The number of tubes varies in both depth and height. Usually one to twelve rows in the direction of airflow (depth) and 4 to 36 tubes per row (height). The more tubes, the more heat transfer, but also the more resistance to airflow and initial cost of the coil. Tube diameters are usually 5/8".

Two-Pipe Direct Return: Two-pipe arrangements have two mains, one for supply and one for return. Each terminal is connected by a supply and return branch to its main. This design not only allows separate control and servicing of each terminal, but because the supply water temperature is the same at each terminal, two-pipe systems can be used for any size application. Two-pipe systems are further distinguished by their return piping. In a two-pipe direct return system the return is routed to bring the water back to the pump by the shortest possible path. The terminals are piped, "first in, first back; last in, last back." The direct return arrangement is popular because generally, less main pipe is needed. However, since water will follow the path of least resistance the terminals closest to the pump will tend to receive too much water while the terminals farthest from the pump will starve. To compensate for this, balancing valves are required.

Two-Pipe Reverse Return: In a two-pipe reverse return system the return is routed so the length of the circuit to each terminal and back to the pump is essentially equal. The terminals are piped, "first in, last back; last in, first back." Because all the circuits are essentially the same length, reverse return systems generally need more piping than direct return systems but are considered more easily balanced. However, balancing valves are still required. Reverse return systems aren't self-balancing. Only in a case where all the terminals were the same size, with the same gpm required, and all piped with identical runouts could a reverse return system be considered self-balancing.

Two-Way Automatic Control Valves: Two-way ACV's control flow rate. They may be either single seated or double seated (balanced valve). The single seated valve is the most common. The valve must be installed with the direction of flow opposing the closing action of the valve plug. The water pressure tends to push the valve plug open. If the valve is installed the opposite way, it'll cause chattering. To understand why this is so, it's important to note that as the valve plug modulates to the closed position, the velocity of the water around the plug becomes very high. Therefore, if the flow and pressure were with the closing of the plug, then at some point near closing the velocity pressure would overcome the spring resistance and force the plug closed. Then, when flow is stopped the velocity pressure goes to zero and the spring takes over and opens the plug. The cycle is repeated and chattering is the result.

The double seated or balanced valve is generally recommended when high differential pressures are encountered and tight shutoff isn't required. The flow through this valve tends to close one port while opening the other port. This design creates a balanced thrust condition which enables the valve to close off smoothly without water hammer despite the high differential pressure.

Valve: Valves are used in hydronic systems for regulating water flow and isolating part, or all the system.

Venturi: A Venturi operates on the same principle as the orifice plate but its shape allows gradual changes in velocity and the "permanent" pressure loss is less than is created by an orifice plate. Calibration data which show flow rate in gallons per minute (gpm) versus measured pressure drop are furnished with the Venturi. The pressure drop is measured with a differential gage.

Waterlogged Compression Tank: The tank becomes waterlogged when the air in the compression tank leaks out and is replaced by water. When this occurs the compression tank can't maintain the proper pressure to accommodate the fluctuations in water volume and control pressure change in the system. A waterlogged tank must be drained and the air leaks found and sealed.

Chapter 16
Fan, Pump, Drive, Motor And Electrical Terminology

FAN TERMINOLOGY

Air Horsepower (AHP): The theoretical horsepower required to drive a fan if the fan were 100% efficient.

Brake Horsepower (BHP): The actual power required to drive a fan.

Clockwise Rotation(CW): As viewed from the drive side of a centrifugal fan.

Counterclockwise Rotation (CCW): As viewed from the drive side of a centrifugal fan.

Cutoff Plate: The cutoff plate is part of a centrifugal fan housing. It's found in the fan outlet. If the cutoff plate isn't properly positioned, air will be drawn back into the fan wheel, losing efficiency.

Double Inlet-Double Width (DIDW): DIDW fans have two single wide fan wheels mounted back-to-back on a common shaft in a single housing. Air enters both sides of the fan. On DIDW fans the bearings are in the air stream and they are large in relation to the inlet and will reduce fan performance if the fan is small. Therefore, DIDW fans are less common in smaller sizes. Generally, because of the double inlet, DIDW fans are more suited to open inlet plenums. They're used most often in high volume applications.

Drive Side: On SISW fans the drive side is the side opposite the inlet. On DIDW fans, the drive side is the side that has the drive.

Fan: A power-driven, constant volume machine which moves a continuous flow of air by converting rotational mechanical energy to increase the total pressure of the moving air.

Fan Air Volume: The rate of flow expressed at the fan inlet in cubic feet per minute (cfm) of air produced, independent of air density.

Fan Blast Area: The fan outlet area less the area of the cutoff.

Fan Efficiency: The output of useful energy divided by the power input; air horsepower divided by brake horsepower.

Fan Outlet Area: The gross inside area of the fan outlet expressed in square feet.

Fan Outlet Velocity (OV): The theoretical velocity of the air as it leaves the fan outlet. It's calculated by dividing the air volume by the fan outlet area. Since all fans have a non-uniform outlet velocity, the "fan outlet velocity" doesn't express the velocity conditions that exist at the fan outlet but is a theoretical value of the velocity that would exist in the fan outlet if the velocity were uniform.

Fan Static Efficiency (FSE): Static air horsepower divided by brake horsepower.

Fan Static Pressure (FSP): The fan total pressure less the fan velocity pressure. The fan inlet velocity head is assumed equal to zero for fan rating purposes.

Fan Total Efficiency (FTE): Total air horsepower divided by brake horsepower.

Fan Total Pressure (FTP): The rise in total pressure from the fan inlet to the fan outlet; the measure of total mechanical energy added to the air by the fan.

Fan Total Static Pressure (TSP): The static pressure rise across the fan calculated from static pressure measurements at the fan inlet and outlet.

Fan Velocity Pressure (FVP): The pressure corresponding to the average air velocity at the fan outlet.

Non-Overloading Fan: The horsepower curve increases with an increase in air quantity but only to a point to the right of maximum efficiency and then gradually decreases. If a motor is selected to handle the maximum brake horsepower shown on the performance curve, it won't be overloaded in any condition of fan operation. Backward curved and backward inclined fans are "non-overloading" fans.

Performance Curve: A fan performance curve is a graphic representation of the performance of a fan from free delivery to no delivery.

Single Inlet-Single Width (SISW): SISW fans have one fan wheel and a single entry. They're more suited to having inlet duct attached to it than DIDW fans. The bearings are out of the air stream.

Static Pressure (SP): The pressure or force within the unit or duct that exerts pressure against all the walls and moves the air through the system. It's sometimes called "bursting pressure."

Tip Speed (TS): The velocity in feet per minute at the tip of the fan blade.

Total Pressure (TP): The sum of the static pressure and the velocity pressure taken at a given point of measurement.

Velocity Pressure (VP): Velocity pressure is the pressure caused by the air being in motion and has a direct mathematical relation to the velocity of the air. Velocity pressure can't be measured directly as can static and total pressure. However, since total pressure is the sum of static and velocity pressure (TP = SP + VP) then velocity can be determined by subtracting static pressure from total pressure (VP = TP − SP).

PUMP TERMINOLOGY

All-Bronze Pumps: All-bronze pumps are used mainly for pumping high temperature water, caustics, sea water, brines, etc. These pumps have the volute, impeller and all parts of the pump coming in direct contact with the liquid made of bronze. The shaft has a bronze sleeve.

All-Iron Pumps: All-iron pumps are used for pumping caustics, petroleum products, etc. As the name implies, all parts of the pump coming in direct contact with the liquid pumped are made of iron or ferrous material.

Axial Flow Impeller: Axial flow impellers have propeller vanes. Speed range: 7,500 to 14,000 rpm.

Bearings: Pumps are equipped with various types of bearings—sleeve, roller or ball. Ball bearings are most common. On double suction pumps, the bearings are located on both sides of the pump casing while on single suction pumps the bearings are located between the pump and the motor.

Belt Driven Pump: A pump equipped with a pulley on the pump shaft and connected by belt to a pulley on the motor shaft.

Brake Horsepower (BHP): The actual power required to drive a pump. Water horsepower divided by pump efficiency.

Bronze-Fitted Pumps: These pumps have a cast iron casing and are equipped with a brass impeller. The metal parts of the seal assembly are made of brass or some other non-ferrous meterial. The shaft is steel. Bronze-fitted pumps are used for most hydronic applications.

Cavitation: The phenomena occurring in a flowing liquid when the pressure falls below the vapor pressure of the liquid, causing the liquid to vaporize and form bubbles. The bubbles are entrained in the flowing liquid and are carried through the pump impeller inlet to a zone of higher pressure where they suddenly collapse or implode with terrific force. The following are symptoms of a cavitating pump: snapping and crackling noises at the pump inlet, severe vibration, a drop in pressure and brake horsepower, and a reduction in flow, or the flow stops completely.

Closed Impeller: An impeller having shrouds or side walls. Designed primarily for handling clear liquids, such as water.

Couplings: Shaft couplings compensate for small deviations in alignment between the pump and motor shafts within the tolerances established by the manufacturer. Couplings are made in "halves" so the pump and motor may be disconnected from each other. It's important that shafts be aligned as closely as possible for quiet operation and the least coupling and bearing wear. The coupling can accommodate small variations in alignment, but its function is coupling, not compensating for misalignment. Severe misalignment between the shafts will lead to noisy operation, early coupling failure, and possible pump or motor bearing failures.

Diffusion Vane or Turbine Pump: A pump built with a series of guide vanes or blades around the impeller. The diffusion vanes have small openings near the impeller and enlarge gradually to their outer diameter where the liquid flows into a chamber and around to the pump discharge.

Direct Drive Pump: A pump mounted on a baseplate and direct connected to its driver motor through a flexible coupling. Most hydronic pumps are direct drive.

Double Suction Pump: A pump in which the liquid enters the impeller inlet from both sides. The impeller is similar to two single suction impellers, back to back. Double suction pumps have fixed suction and discharge openings. The suction connections are normally one or two pipe sizes larger than the discharge connection.

Dynamic Discharge Head: Static discharge head plus friction head plus velocity head.

Dynamic Suction Head: Static suction head minus friction head loss and velocity head.

Dynamic Suction Lift: Static suction lift plus friction head loss plus velocity head.

Equalized Spring Coupling: The equalized spring type coupling is used where quiet, smooth operation is required. The motor drives the pump shaft through four springs. The tension on the springs is balanced by an equalizing bar. This coupling needs no maintenance when the alignment is proper. Check alignment if the coupling breaks or noisy operation is observed.

Flexible Disc Coupling: The flexible disc is used in heavy duty applications where extremely quiet conditions aren't needed. The motor drives the pump shaft through the flexible disc. If rough operation or noise is observed, check the coupling for wear. The flexible disc should never be tightly bound between the two coupler halves. Check with the manufacturer for clearance specifications.

Friction Head: The pressure required to overcome the resistance to flow, expressed in psi or feet of head.

Mechanical Seals: Mechanical seals have a stationary ring, usually made of hard ceramic material, and a rotating graphite ring. The stationary ring fits into a recess in the pump body and has a rubber gasket behind it which forms a water-tight seal. Behind the molded graphite ring, is a rubber bellows and a seal spring. It's this spring that keeps the rotating graphite ring tight against the face of the ceramic ring making the water seal. No maintainance or adjustments are needed as the spring continually pushes the graphite ring forward to make up for wear. It's important that a water film is between the two surfaces to provide lubrication and cooling. Running or bumping a pump, without water in the system will damage mechanical seals. When the seal leaks, replace it.

Mixed Flow Impeller: Some mixed flow impellers use double curvature vanes. These are wider than the plain flow impeller. Speed range: 2,500 to 5,000 rpm. Other mixed flow impellers use propeller vanes. Speed range: 5,000 to 9,000 rpm.

Multi-Stage Pump: A pump having two or more impellers on a common shaft, acting in series in a single casing. The liquid is conducted from the discharge of the preceding impeller through fixed guide vanes to the suction of the following impeller causing a head (pressure) increase at a given flow rate as it passes through each stage. Head can also be increased by connecting separate single stage pumps in series. Multi-stage pumps have been built with over 300 stages.

Net Positive Suction Head (NPSH): The minimum suction pressure at the pump to overcome all the factors limiting the suction side of the pump— internal losses, elevation of the suction supply, friction losses, vapor pressure and altitude of the installation. It's unlikely there will be a problem with NPSH in air conditioning chilled water closed circuit systems. It's also not ordinarily a factor in open systems or hot water systems unless there's considerable friction loss in the pipe or the water source is well below the pump and the suction lift is excessive. If there's insufficient net positive suction head available (NPSHA) check the suction line for undersized pipe, too many fittings, throttled valves, or clogged strainers.

Net Positive Suction Head Available (NPSHA): NPSHA is a characteristic of the system in which the pump operates. It's dependent upon such conditions as elevation of the suction supply in relation to the pump, the friction loss in the suction pipe, the altitude of the installation or the pressure on the suction supply, and vapor pressure. In determining the NPSHA, these considerations must be evaluated and a pump selected for the worst conditions likely to be encountered in the installation. In addition, as a safety factor, the NPSHA should always exceed the NPSHR by two feet or more.

Net Positive Suction Head Required (NPSHR): NPSHR is the actual absolute pressure needed to overcome the pump's internal losses and allow the pump to operate satisfactorily. NPSHR is determined by the pump manufacturer through laboratory tests. It's a fixed value for a given capacity and doesn't vary with altitude or temperature. It does vary with each pump capacity and speed change. NPSHR for a specific pump is available from the manufacturer either on submittal data, from a pump curve or from a catalog. A pump curve will give the full range of NPSHR values for each impeller size and capacity.

Non-Overloading Pump: The horsepower curve increases with an increase in water quantity but only to a point near maximum capacity, and then gradually decreases. If a motor is selected to handle the maximum brake horsepower shown on the performance curve, it won't be overloaded in any condition of pump operation.

Open Impeller: An impeller without side walls, consisting essentially of a series of vanes attached to a central hub. Used for handling liquids containing abrasives or solids.

Performance Curve: A pump performance curve is a graphic representation of the performance of a pump. Pumps are generally selected so their design operating point falls about midway, plus or minus ¼, of the published curve. This allows changes in installation conditions.

Primary-Secondary Pumps: The function of the primary pump in a primary-secondary circuit is to circulate water around the primary circuit. The function of the secondary pump is to supply the terminals.

Pump: A machine for imparting energy to a fluid. The addition of energy to a fluid makes it do work such as rising to a higher level or causing it to flow. In a hydronic system, the pump is the component which provides the energy to overcome system resistance and produce the required flow.

Pump Efficiency: The output of useful energy divided by the power input; water horsepower divided by brake horsepower.

Radial Flow Impeller: The radial or plain flow impeller is the most frequently used for HVAC work. This impeller design has single curvature vanes which curve backwards and are used in pumps with speed ranges below 2,500 rpm.

Semi-Open Impeller: An impeller having a shroud or side wall on one side only, usually on the back. Used for handling liquids containing solids.

Shut-Off Head: The pressure developed by the pump when its discharge valve is shut. On the pump curve, it's the intersection of the head-capacity curve with the zero capacity line.

Single Stage Pump: A pump with one impeller.

Single Suction Pump: A pump in which the liquid enters the impeller inlet from one side. Single suction pumps are usually built with the inlet at the end of the impeller shaft. The casing is made so the discharge may be rotated to various positions. The suction connection is normally one or two pipe sizes larger than the discharge connection.

Specific Gravity (Sp. Gr.): The ratio of the mass of a substance to the mass of an equal volume of water at 4 degrees C. Water at standard conditions has a specific gravity of 1.0.

Static Head: The static pressure of a fluid expressed in terms of the height of a column of the fluid.

Static Discharge Head: The vertical distance from the centerline of the pump to the free discharge liquid level.

Static Suction Head: The vertical distance from the centerline of the pump to the suction liquid free level.

Static Suction Lift: The vertical distance from the centerline of the pump *down* to the suction liquid free level.

Stuffing Box: The stuffing box seal has a "packing" which has rings made of graphite-impregnated cord, molded lead foil or some other resilient material formed into fitted split rings. These packing rings are compressed into the stuffing box by a packing gland. The tension on the packing gland is critical to the proper operation of the pump. If a packing gland must be replaced, consult the manufacturer's published data for tension recommendations. If there's too much tension, the proper water leakage won't occur and this will cause scoring of the shaft and overheating of the packing. Another problem is that as the seal gets older and the packing gland has been tightened over time, the packing becomes compressed and loses its resiliency, overheating the stuffing box. When the packing gland is backed off to allow cooler operation, there's excessive leakage. When this happens replace the packing.

Suction Head: When the source of supply is above the pump centerline.

Suction Lift: When the source of supply is below the pump centerline.

Total Discharge Head: The static discharge head plus friction losses plus velocity head.

Total Dynamic Head (TDH): (1) The total discharge head minus the total suction head or (2) the total discharge head plus suction lift. Suction head is when the water source is above the pump centerline. Suction lift is when the water source is below the pump centerline. For test and balance purposes, TDH is the difference between the gage pressure at the pump discharge and the gage pressure at the pump suction.

Total Head: In a flowing fluid, the sum of the static and velocity heads at the point of measurement.

Total Static Head: The vertical distance in feet from the suction liquid level to the discharge liquid level. The sum of static suction lift and static discharge head. The difference between static suction head and static discharge head.

Vapor Pressure: The vapor pressure of a liquid at any given temperature is that pressure necessary to keep the liquid from boiling or flashing into a vapor. For example, the vapor pressure of 60-degree water is 0.59 feet absolute (0.25 psia) while water at 180 degrees has a vapor pressure of 17.85 feet absolute (7.72 psia).

Velocity Head: The head required to create flow. The height of the fluid equivalent to its velocity pressure.

Volute Pump: A pump having a casing made as a spiral or volute curve. The volute casing starts with a small cross-sectional area near the impeller and increases gradually to the pump discharge.

Water Horsepower (WHP): The theoretical horsepower required to drive a pump if the pump were 100% efficient.

DRIVE TERMINOLOGY

Adjustable Sheave: The belt grooves are adjustable. Adjustable groove sheaves are also known as variable speed or variable pitch sheaves.

Fan Sheave: The driven pulley on the fan shaft.

Fixed Sheave: The belt grooves are fixed. Fixed sheaves are normally used on the fan. Size for size, fixed sheaves are less expensive than variable pitch sheaves and there's less wear on the belts.

Matched Belts: A set of belts whose exact lengths and tensions are measured and matched by the supplier in order for each belt to carry its proportionate share of the load.

Motor Sheave: The driver pulley on the motor shaft. The motor sheave may be either a fixed or adjustable groove sheave. Generally, after fans have been adjusted for the proper airflow, variable pitch motor sheaves are replaced with fixed sheaves.

Pitch Diameter (Pd): Approximately where the middle of the belt rides in the sheave groove.

Shaft: Motor shaft sizes are in 1/8″ increments and fan shafts are in 1/16″ increments.

V-Belts: Two types of V-belts are generally used on HVAC equipment—light duty, fractional horsepower (FHP) belt, sizes 2L through 5L and industrial belts, sizes "A" through "E." Fractional horsepower belts are generally used on smaller diameter sheaves because they're more flexible than the industrial belt for the same equivalent cross-sectional size. For example, a 5L belt and a "B" belt have the same cross-sectional dimension, but because of its greater flexibility, the 5L belt would be used on light duty fans that have smaller sheaves. The general practice in HVAC design is to use belts of smaller cross-sectional size with smaller sheaves instead of large belts and large sheaves for the drive components. Multiple belts are used to avoid excessive belt stress. V-belts are rated by horsepower per belt, by length and minimum recommended pitch diameter.

MOTOR TERMINOLOGY

Brake Horsepower (BHP): The total horsepower applied to the drive shaft of any piece of rotating equipment.

Controllers: Controllers are used for starting and stopping industrial motors. They can be grouped into three categories, manual starters, contactors, and magnetic starters.

Contactors: Electro-mechanical devices that "open" or "close" contacts to control motors.

Dual Current Motor: A motor that will operate safely at either of two nameplate amperages. The operating amperage depends on the voltage supplied.

Dual Voltage Motor: A motor that will operate safely from either of two nameplate voltages. Typical single-phase dual voltage motors are 110/220 volts or 115/230 volts. Typical three-phase motor dual voltage motors are 220/440 volts, 230/460 volts, or 240/480 volts.

Heaters: See overload protection.

Horsepower (HP): A unit of power. One horsepower equals 746 watts.

Locked Rotor Amperage (LRA): Locked rotor amperage occurs between zero and full motor speed when the starting current is drawn from the line with the rotor locked and with rated voltage supplied to the motor. During this short time, a fraction of a second for small motors to a second or longer for large motors, the locked rotor amperage far exceeds the full load operating current. Locked rotor amperage will generally be 5 to 6 times the full load amperage. This inrush of current will continue to decrease until the motor reaches full operating speed.

Magnetic Starters: Contactors with overload protection relays. Sometimes called "mags" or "mag starters."

Manual Starters: Motor-rated switches that have provisions for overload protection. Generally, limited to motors of 10 horsepower
or less.

Nameplate Amps or Full Load Amps (A, FLA): The full load
operating current at rated voltage and horsepower.

Nameplate RPM: The rated motor speed. Motors operate at different speeds according to their type, construction and the number
of magnetic poles in the motor. For instance, a four-pole synchronous motor would be rated at 1800 rpm, but because of
slip, about 2 to 5%, a four-pole induction motor is rated at
1725 rpm.

   Some single-phase motors are designed for multiple rpm by
switching the winding connections so two to four different
speeds are available. Wiring diagrams are usually provided on the
motor.

Nameplate Volts (V): The rated operating voltage.

Overload Protection (OL): Thermal overload protection devices,
sometimes called "heaters" or "thermals" prevent motors from
overheating. If a motor becomes overloaded or one phase of a
three-phase circuit fails (single phasing), there will be an increase
in current through the motor. If this increased current drawn
through the motor lasts for any appreciable time and it's greatly
above the full load current rating, the windings will overheat and
damage may occur to the insulation, resulting in a burned-out
motor. Because most motors experience various load conditions
from no load to partial load to full load to short periods of being
overloaded, their overload protection devices must be flexible
enough to handle the various conditions under which the motor
and its driven equipment operate. Single-phase motors often
have internal thermal overload protection. This device senses the
increased heat load and breaks the circuit, stopping the motor.
After the thermal overload relays have cooled down, a manual
or automatic reset is used to restart the motor. Other single-phase
and three-phase motors require external overload protection.

Service Factor (SF): The number by which the horsepower or amperage rating is multiplied to determine the maximum safe load
that a motor may be expected to carry continuously at its rated
voltage and frequency. Typical service factors are 1.0, 1.10, 1.15

for large motors, and 1.20, 1.25, 1.30 and 1.40 for small motors.

Single Phasing: The condition which results when one phase of a three-phase motor circuit is broken or opened. Motors won't start under this condition but if already running when it goes into single-phase condition, the motor will continue to run with a lower power output and possible overheating.

Single-Phase Motor (1φ): A motor supplied with single-phase current.

Three-Phase Motor (3φ): A motor supplied with three-phase current. For the same size, three-phase motors have a capacity of about 150% greater, are lower in first cost, require less maintenance and generally do better than single-phase motors.

Thermals: See overload protection.

Torque: The force which produces or tends to produce rotation. Measured in foot-pounds (ft-lb).

ELECTRICAL TERMINOLOGY

Alternating Current (AC): The type of electrical circuit in which the current constantly reverses flow. The standard electrical service used in HVAC systems is single-phase or three-phase alternating current at 60 cycles per second.

Amerpage: Electron (current) flow of one coulomb per second past a given point.

Ampere, AMP (A): A measure of electrical flow rate. One ampere is equal to one coulomb per second or 6.3×10^{18} electrons per second.

Conductor: A material which readily passes electrons. Good conductors are silver, copper and aliminum wire. Copper, the second best conductor, is the most often used in electrical wiring because of price and availability.

Current (I): The transfer of electrical energy through a conductor.

Direct Current (DC): The type of electrical circuit in which the current always flows in one direction.

Electric: Derived from the Latin word "electricus" meaning amber like. Amber is a hard, translucent, yellow fossil resin found along the shores of the Baltic Sea. Amber, when rubbed with a cloth gives off sparks. The ancient Greeks used words meaning "electric force" in referring to the mysterious forces of attraction and repulsion exhibited by amber.

Electricity: The effect created by the interaction of charged particles. Electricity is defined only by theory. However, these theories about the nature and behavior of electricity have been advanced, and have gained wide acceptance because of their apparent truth and demonstrated workability. The theory of electricity or electron flow states that negatively charged particles, called electrons, flow from point to point through a conductor.

Electromotive Force (EMF): A measure of electric force or potential voltage.

Energy: A measure of power consumed (energy = power x time). Measured in watt-hours.

Frequency (F): The number of complete cycles per second (cps) of alternating current.

Ground (GND): The earth. The lowest potential or voltage of an electrical system.

Hertz (Hz): The number of complete cycles per second of alternating current.

Hot Wire: Any wire which is at a higher voltage than neutral or ground. Also, leg or lead.

Impedance (Z): The total opposition to the flow of alternating current by any combination of resistance, inductance and capacitance. Impedance is measured in Ohms.

Insulator: A material which doesn't readily pass electrons. Some of the best insulators are rubber, glass and plastic.

Induction: The process of producing electron flow by the relative motion of a magnetic field across a conductor. In a transformer, current flowing through the primary coil sets up a magnetic field. This magnetic field produces current flow in the secondary coil.

Kilovolt-Ampere (KVA): 1000 volt-amperes.

Kilowatt (kW): 1000 watts.

Kilowatt-Hour (kWh): 1000 watt-hours.

Neutral (N): That part of an electrical system which is at zero voltage difference with respect to the earth or "ground."

OHM (R): A measure of resistance in an electrical circuit.

Open Circuit: The condition which exists when either deliberately or accidentally an electrical conductor or connection is broken or opened with a switch.

Phase (PH ϕ): The number of separate highest voltages alternating at different intervals in the circuit.

Power (P): The rate of doing work. Electrical power is measured in watts or kilowatts. Other units of power are horsepower and Btuh.

Power Factor (PF): The ratio of actual power to apparent power.

Short Circuit: The condition which occurs when a hot wire comes in contact with neutral or ground.

Volt or Voltage (V or E): A measure of electric force or potential. Also called electromotive force.

Volt-Ampere (VA): A unit of apparent power. Volts times amperes.

Watt (W): A unit of actual power. For resistance loads such as strip heaters: $W = VA$. For induction loads such as AC motors: $W = VA \times PF$.

Watt-Hour (Whr): A measure of energy. Watts times hours.

Chapter 17
Instruments
And Instrument Terminology

CARE AND USE OF INSTRUMENTS

1. Protect instruments against shock, vibration and temperature.

2. Use only calibrated test instruments to take readings. Permanently installed thermometers and pressure gages deteriorate with age and aren't reliable.

3. Field instruments should be checked for calibration against a sheltered set of instruments before each use.

4. Follow the instrument manufacturer's recommended procedures.

Air Instruments
Allow ample time for instruments to stabilize to ambient conditions.

Electrical Instruments
Continue reading until two repeatable consecutive values are obtained.

Rotational Speed Instruments
Continue reading until two repeatable consecutive values are obtained.

Temperature Instruments
1. Allow ample time for the temperatures to stabilize.

2. Where applicable, use the same thermometer for all readings.

3. Continue reading until two repeatable consecutive values are obtained.

Water Instruments
1. Allow ample time for instruments to stabilize to ambient conditions.

2. Drain all water from instruments after use (this is especially criti-
    cal in cold climates).

COMMON TERMS

Absolute Pressure: The total of the indicated gage pressure plus the
atmospheric pressure. Abbreviated "psia" for pounds per square
inch absolute.

Atmospheric Pressure: The pressure exerted upon the earth's surface
by the air because of the gravitational attraction of the earth.
Standard atmosphere pressure at sea level is 14.7 pounds per
square inch (psi). Measured with a barometer.

Barometer: An instrument for measuring atmospheric pressure.

Calibration: Determining or correcting the error of an existing scale.

CEFAPP: Close enough for all practical purposes.

Differential Pressure Gage: An instrument that reads the difference
between two pressures directly and therefore, eliminates the need
to take two separate pressures and then calculate the difference.

Electronic Instruments: Most of the mechanical analog instruments
now have electronic digital counterparts. All instruments, analog
or digital, should be checked against a sheltered set before each
balancing project. Pressure measuring instruments should be
checked against a standard liquid-filled manometer.

Gage: An instrument for measuring pressure.

Gage Pressure: The pressure that's indicated on the gage.

Harmonics: For a strobelight tachometer, harmonics are frequencies
of light flashes that are a multiple or submultiple of the actual
rotating speed. For example, if the light frequency is either
exactly two times or exactly one-half the actual speed of the
rotating equipment, the part will appear stationary but the image
won't be as sharp as when the rpm is correct.

Manometer: An instrument for measuring pressures. Essentially a
U-tube partly filled with a liquid, usually water, mercury or a
light oil. The pressure exerted on the liquid is indicated by the
liquid displaced. A manometer can be used as a differential
pressure gage.

Meniscus: The curved surface of the liquid column in a manometer. In manometers that measure air pressures, the liquid is either water or a light oil. In manometers that measure water pressures, the liquid is mercury.

Operating Load Point: Actual system operating capacity when an instrument reading is taken.

Parallax: A false reading that happens when the eye of the reader isn't exactly perpendicular to the lines on the instrument scale.

Pitot Tube: A sensing device used to measure total pressures in a fluid stream. It was invented by a French physicist, Henri Pitot, in the 1700's.

Sheltered Set: A sheltered set of instruments is a group of instruments used only to check the calibration of field instruments.

Sensitivity: A measure of the smallest incremental change to which an instrument can respond.

AIR INSTRUMENTS

Air Differential Pressure Gage: Magnehelic® and Capsuhelic® are two brands of air differential gages. These gages contain no liquid, but instead work on a diaphragm and pointer system which move within a certain pressure range. Although absolutely level mounting isn't necessary, if the position of the gage is changed, resetting of the zero adjustment may be required for proper gage reading as specified by the manufacturer. These gages have two sets of tubing connector ports for different permanent mounting positions; however, only one set of ports is used for readings and the other set is capped off. The ports are stamped on the gages as "high" pressure and "low" pressure and when used with a Pitot tube, static tip or other sensing device to measure total pressure, static pressure, or velocity pressure, the tubing is connected from the sensing device to the tubing connector ports in the same manner as the connection to the inclined-vertical manometer. These gages are generally used more for static pressure readings than for velocity pressures readings. They should be checked frequently against a manometer.

Anemometer: An instrument used to measure air velocities.

Capture Hood: An instrument which captures the air of a supply, return or exhaust terminal and guides it over a flow-measuring device. It measures airflow directly in cubic feet per minute. Calibration by the manufacturer should be done every 6 months, especially if the instrument isn't checked periodically against a sheltered instrument.

Compound Gages: Compound gages measure pressures both above and below atmospheric. They read in pounds per square inch above atmospheric and inches of mercury below atmospheric. Compound gages are calibrated to read zero at atmospheric pressure.

Cubic Feet Per Minute (CFM): A unit of measurement. The volume or rate of airflow.

Deflecting Vane Anemometer: The deflecting vane anemometer gives instantaneous, direct readings in feet per minute. It's used most often for determining air velocity through supply, return and exhaust air grilles, registers or diffusers. It may also have attachments for measuring low velocities in an open space or at the face of a fume hood. With other attachments, Pitot traverses and static pressures can be taken. To use this instrument refer to the manufacturer's recommendation for usage, proper attachment selection and sensor placement. A correction (Ak) factor is also needed when measuring grilles, registers or diffusers. Calibration by the manufacturer should be done every 6 months, especially if the instrument isn't checked periodically against a sheltered instrument.

Feet Per Minute (FPM): A unit of measurement. The velocity of the air.

Hot Wire Anemometer: This instrument measures instantaneous air velocity in feet per minute using an electrically heated wire. As air passes over the wire, the wire's resistance is changed and this change is shown as velocity on the instrument's scale. This instrument is very position sensitive when used to measure air velocities. Therefore, it's important to ensure that the probe is held at right angles to the airflow. The hot wire anemometer is most often used to measure low velocities such as found at the face of fume hoods; however, some instruments can also measure temperatures and static pressures. Calibration by the manufacturer should be done every 6 months, especially if the instrument isn't checked periodically against a sheltered instrument.

Inches of Water Gage or Column (IN. WG or IN. WC): A unit of air pressure measurement equal to the pressure exerted by a column of water 1 inch high.

Inclined Manometer, Inclined-Vertical Manometer: The inclined manometer has an inclined scale which reads in inches of water gage in various ranges such as 0 to 0.25 in.wg, 0 to 0.50 in. wg, or 0 to 1.0 in. wg. The inclined-vertical manometer has both an inclined scale that reads 0 to 1.0 in. wg and a vertical scale for reading greater pressures such as 1.0 to 10 in. wg.

The inclined and inclined-vertical manometer have a left and right tube connection for attaching tubing from a Pitot tube or other sensing device to the manometer. The left tubing connector is called the "high" side of the manometer and the right connector is the "low" side. The manometers are filled with a colored oil which is lighter than water and, therefore, the oil will move a greater distance for a given pressure allowing more precise readings.

Magnehelic® Gage: An instrument for measuring air pressures and differential pressures. See air differential pressure gage.

Micromanometer: Micromanometers are instruments generally suited more to laboratory testing than to field measurements. The micromanometer is difficult to use in the field because of the leveling and mounting requirements. Micromanometers are used to measure very low pressures accurately down to plus or minus one thousandth (0.001) inch of water gage. If field measurements below 0.03 in. wg (700 fpm) are required, consider using a 0. to 0.25" inclined manometer, a micromanometer or hot wire anemometer.

Manometer: The manometer is the standard of the industry for reading air pressure. It contains no mechanical linkage and doesn't need calibration. Other pressure measuring instruments are checked for calibration against a properly set up and accurately read manometer. Types of manometers commonly used to measure air pressures are the inclined manometer, the inclined-vertical manometer, the U-tube manometer and the micromanometer.

Pitot Tube: The standard Pitot tube has a double tube construction with a 90-degree radius bend near the tip and measures both total and static pressures. It is 5/16" in diameter and is available in various lengths from 12" to 60". The standard tube is recommended for use in duct 8" and larger diameter while a smaller (1/8" diameter) "pocket" Pitot tube is used in ducts smaller than 8". To help in taking Pitot tube traverses, the outer tube is marked with a stamped number at the even-inch points and a 1/8" line at the odd-inch intervals. The inner tube, or impact tube, senses total pressure and runs the length of the Pitot tube to the total pressure connection at the bottom.

Static pressure is sensed by the outside tube through eight equally spaced holes around the circumference of the tube. These small openings must be kept clean and open to have accurate readings. They're located near the tip of the Pitot tube. The air space between the inner and outer tube serves to transmit the static pressure from the sensing holes to the static pressure connection at the bottom side of the Pitot tube. The static pressure connection is parallel with the tip of the Pitot tube as an aid to aligning the tip properly.

To ensure the accurate sensing of pressures, the Pitot tube tip must be pointed so it faces directly into and parallel with the air stream. The hookups of the Pitot tube to an inclined or inclined-vertical manometer or air differential pressure gage (ADPG) are

1. Total Pressure (TP)
   a. Open the high and low tubing connectors on the manometer or the ADPG.
   b. Connect one piece of tubing to the total pressure connection on the Pitot tube.
   c. If the total pressure reading is on the discharge of the fan, connect the other end of the tubing to the left (high side) tubing connector on the manometer (high port of the ADPG). If the total pressure reading is on the inlet of the fan, connect the tubing to the right (low side) tubing connector (low port of the ADPG).
2. Static Pressure (SP)
   a. Open the high and low tubing connectors on the man-

ometer or the ADPG.

b. Connect one piece of tubing to the static pressure connection on the Pitot tube.

c. If the static pressure reading is on the discharge side of the fan, the other end of the tubing is connected to the left (high side) tubing connector on the manometer (high port of the ADPG). If the static pressure reading is on the inlet of the fan, the tubing is connected to the right (low side) tubing connector (low port of the ADPG).

The tubing connections to the manometer or ADPG are reversed when taking total and static pressure because the air pressures on the discharge of the fan are greater than atmospheric pressure whereas, on the inlet of the fan, the air pressures are less than atmospheric. If the tubing isn't connected to the proper side of the manometer, not only will the reading be incorrect, but also there's a good chance that oil will be blown out of the manometer.

3. Velocity Pressure (VP)

a. Open the high and low tubing connectors on the manometer or the ADPG.

b. Using two pieces of tubing, connect one piece of tubing to the total pressure connection on the Pitot tube and the other piece of tubing to the static pressure connection.

c. Connect the total pressure tubing to the left (high) side of the manometer (high port of the ADPG).

d. Connect the static pressure tubing to the right (low) side of the manometer (low port of the ADPG).

Velocity pressure is the subtraction of static pressure from total pressure and because total pressure is always greater than or equal to static pressure, velocity pressure will always be a positive value. This means that when measuring velocity pressure using the Pitot tube, the hookup is always the same no matter if the reading is taken on the discharge of the fan or on the inlet.

Rotating Vane Anemometer (RVA): Using the rotating vane anemometer for proportional balancing is satisfactory when properly applied; however, where applicable, actual flow quantities should be verified by a Pitot tube traverse.

The RVA is generally used for determining air velocity through supply, return and exhaust air grilles, registers or openings. It's also sometimes used to measure total airflow through coils. However, it isn't a recommended practice to measure airflow through coils or dampered registers since the air through these devices exists in thin, high-speed jets. The rotating vane anemometer is not an averaging instrument and if such measurements are attempted, it'll provide a false (high) reading.

A rotating vane anemometer measures the linear feet of the air passing through it. Because the RVA reads in feet, a stopwatch or other timing device must be used to find velocity in feet per minute. The useful velocity range of the RVA is between 200 to 2000 fpm and the accuracy of the instrument depends on the precision of use, the type of application, and its calibration. Calibration should be checked frequently against a manometer. Also, calibration by the manufacturer should be done every 6 months, especially if the instrument isn't checked periodically against a sheltered instrument. The manufacturer should return the instrument with a velocity correction chart for that instrument.

The RVA (either a 3″, 4″ or 6″ diameter [the 4″ is preferred]) is placed in the air stream with the air flowing from back to front. This means that RVA is reversed for return and exhaust air readings. There's a shroud that fits on the front of the RVA which should be used when taking readings on return and exhaust air grilles to keep the dial face from being scratched. The shroud should be held tightly against the face of the grille.

All readings should be taken with the handle attachment in place to avoid interference with the airflow. All internal dampers on registers must be full open and all adjustable face bars must be at zero degrees deflection. Readings may be taken for 15, 30 or 60 seconds, depending on uniformity and velocity of flow. Follow the manufacturer's recommendations for measuring airflow.

If a manufacturer's Ak isn't available or appears erroneous, establish a new Ak by Pitot tube traverse of the duct. If this isn't practical, and a reading is needed for proportioning purposes, (1) hold the RVA 1″ away from the face of a supply grille or register. This will allow the air to recover from the vena contracta effect of the air passing through the face bars. For a return grille or register, hold the RVA against the face bars; (2) use the inside dimensions of the grille (not the nominal grille size) for calculating area correction factor. Don't subtract the area of face bars from the calculated area factor. If the smallest inside dimension is less than 6", use the diameter of the rotating vane anemometer.

If the airflow is of similar quantities over the face of the opening, grille, etc., a serpentine traverse may be taken. However, if the air velocities differ greatly, the opening should be sectioned off as needed to get uniform flow. The readings are taken with the rotating vane anemometer traversing or stationary in each section.

Example: A supply register is 24" x 12". There' no manufacturer's Ak. The calculated Ak is 1.9 (23.625" x 11.625" the inside dimensions, divided by 144 sq. in. per sq. ft.). The face velocities are dissimilar, 0 fpm in some areas and 500 fpm in others. This register face is divided into four 12" x 6" sections. A reading is taken in each section for 15 seconds. After each timed interval the RVA is stopped and removed from the air stream. The value on the indicating dial is read and the instrument correction factor is applied. This number is multiplied times 4. After the readings in all the sections are taken, the average velocity in feet per minute is calculated and multiplied times the Ak factor to get air volume (CFM = Ak x V).

Standard Cubic Feet Per Minute (SCFM): The volumetric rate of airflow at standard air conditions. See standard air conditions under Psychrometric Terminology, Chapter 13.

U-Tube Manometer: A manometer with a U-shaped glass or plastic tube partly filled with tinted water, or oil. They're made in various sizes and are recommended for measuring pressures of several inches of water gage or more. They're not recommended for readings of less than 1.0 in. wg. To take a reading, open both tubing connectors. The liquid will be at the same height in each leg. A Pitot tube, impact tube or static pressure tip is connected to the manometer. When the sensing device is inserted into the duct or fan compartment the liquid is forced down in one leg and up in the other. The difference between the heights of the two legs is the pressure reading.

ELECTRICAL INSTRUMENTS

Ammeter: An instrument for measuring amperage.

Multimeter: An instrument that measures more than one electrical component. A multimeter may measure AC current and voltage, DC current and voltage, and power factor. Calibration should be every 6 months.

Power Factor Meter: An instrument for measuring power factor. Read all phases from left to right. Also, read the corresponding amperage of each phase and the voltages between each phase. Calibration should be every 6 months.

Voltammeter: A multimeter for measuring voltage and amperage. The portable voltammeter is used most. It may have an analog or digital scale. To prevent pegging the movable pointer when using an instrument with an analog scale, start at the highest scale and work down until the measured voltage or amperage is read in the upper half of the scale. The ammeter part of the instrument has a trigger-operated set of clamp-on transformer jaws which permit the taking of amperage readings without interrupting electrical service.

To take voltage readings on a single-phase motor, measure voltage between phase and ground. On a three-phase motor, measure the voltage between each phase. The standard practice is to read the phase-to-phase voltage from left to right. Read the voltage between the left (T1) and center (T2) line terminals, the left (T1) and right (T3) line terminals, and then the center (T2) and right (T3) line terminals.

To take amperage readings on a three-phase motor, read from left to right (Line 1, Line 2 and Line 3). Only one reading is needed when measuring a single-phase circuit and that reading can be on either the hot wire of the neutral wire. Calibration should be every 6 months.

Voltmeter: An instrument for measuring voltage.

ROTATIONAL SPEED INSTRUMENTS

Chronometric Tachometer: A type of contact tachometer. It's a combination of a precision stop watch and a revolution counter.

Contact Tachometers: Chronometric and digital are two types of contact tachometers. Included with contract tachometers are rubber and metal tips for centering the instrument on the rotating shaft. The rotating shaft should be clean to ensure proper contact with the tip. The tip should be properly centered on the rotating shaft and held against the shaft tight enough to avoid slippage but not so tight as to increase drag in the tachometer to the point of causing an incorrect reading. The instrument shaft should be held parallel to the rotating shaft to ensure an accurate reading. Calibration should be every 6 months.

Non-Contact Tachometers: Non-contact tachometers are used for measuring rotational speeds when the shaft isn't accessible. Two types are the strobelight tachometer and the photo tachometer. Calibration should be every 6 months.

Photo Tachometer: This instrument measures rpm by flashing a light at the moving part and counting the reflections.

Strobelight Tachometer: A strobelight tachometer has an electronically controlled flashing light which is manually adjusted to equal the frequency of the rotating part so the part will appear motionless. To avoid reading harmonics of the actual rpm, use nameplate rpm or calculate the approximate rpm and start at that point. For example, if the speed of a fan needs to be measured with a strobelight, use the drive equation to determine the approximate rpm of the fan.

Tachometer: An instrument for measuring speed.

TEMPERATURE INSTRUMENTS

Bimetallic Thermometer: Bimetallic dial thermometers are acceptable when they're calibrated against a standard glass stem mercury thermometer before each use.

Digital Thermometer: Digital thermometers are acceptable when they're calibrated against a standard glass stem mercury thermometer before each use or are capable of being field calibrated. They should read in the tenths of degrees and be accurate to one-tenth of a degree.

Glass Stem Thermometer: Glass stem thermometers are generally limited to hand-held immersion readings in applications where precise readings are required. These thermometers can also be used as a standard for calibrating and verifying other types of thermometers if the error of the glass stem thermometer doesn't exceed plus or minus one scale division.

Psychrometer: Sling or battery powered psychrometers are used to measure wet-bulb and dry-bulb temperatures. The power psychrometer is preferred. The wick of the wet-bulb thermometer must be kept clean and wetted with distilled water only.

Pyrometer: Pyrometers are generally used to measure pipe surface temperatures. Calibration should be every 6 months.

WATER INSTRUMENTS

Bourdon Tube Test Gages: Bourdon tube "test" gages are high quality Bourdon tube gages with mirrored backs to help eliminate parallax error. They have an accuracy of one-half of one percent of maximum scale reading and they're widely used for test and balance work to measure static pressures at pumps, terminal units, and primary heat exchange equipment such as chillers and condensers. Some test gages are designed to measure pressures above atmospheric (pressure gages) and read in pounds per square inch (psi) while other gages which measure pressures below atmospheric are known as "vacuum gages" and read in inches of mercury (in. Hg). There are test gages that can measure both above and below atmospheric and they're called "compound gages" and read in pounds per square inch above atmospheric and inches of mercury below atmospheric. Bourdon tube pressure test gages, vacuum test gages and compound test gages are available in various ranges but all are calibrated to read zero at atmospheric pressure.

Test gages should be selected so the pressures measured fall in the upper one-half of the scale. Don't subject the test gage to pressures above or below the limits of the scale and use a pulsation snubber, restrictor, or needle valve to stop system pressure pulsations. Also, in using test gages apply and remove pressure slowly by gradually opening the gage cock. If these measures are taken it'll extend the life of the test gage and allow greater accuracy of pressure readings. Calibration should be every 6 months.

When using a Bourdon tube test gage, ensure that the gage is at the same height for both the entering and leaving reading or that a correction is made for a difference in height. The need for correction and the possibility of error resulting from the correction can be eliminated by using only one test gage with a manifold. With a single gage connected in this manner, the gage is alternately valved to the high pressure side and then the low pressure side to determine the pressure differential, eliminating any problem about gage elevation.

Compound Gages: Compound gages measure pressures above and below atmospheric. They read in pounds per square inch above atmospheric and inches of mercury below atmospheric.

Pressure Gages: Measure pressures above atmospheric and read in pounds per square inch, gage (psig). Pressure gages are calibrated to read zero at atmospheric pressure. Permanently installed gages deteriorate because of vibration and pulsation and aren't reliable for test and balance measurements. Use only recently calibrated test gages.

Snubber (Pulsation Suppressor): A snubber is a restrictor placed in the water line to a permanently installed gage to suppress pulsating or fluctuating pressures.

U-Tube Manometers: U-tube manometers are primarily used for measuring pressure drops across terminals, heat exchangers, and flow meters. Fundamentally, the principles that apply to U-tube manometers that measure air pressures also apply to U-tube manometers that measure water pressures. The differences are (1) the pressures measured in a hydronic system are usually much greater than in air systems, (2) the U-tube manometer contains mercury instead of oil or water and must have over-pressure traps or use a manifold to prevent the mercury from entering the piping system (mercury causes rapid deterioration of copper and copper alloy pipes), (3) all air must be purged from the manometer and hoses, (4) water from the system must completely fill the hoses and both manometer legs and rest on top of the mercury.

U-tube manometers may also be used with a water Pitot tube to take a velocity head traverse of the pipe. **Caution:** The high pressures associated with some water systems can drive the Pitot tube back into the user.

Vacuum Gages: Vacuum gages measure pressures below atmospheric and read in inches of mercury (in. Hg). Vacuum gages are calibrated to read zero at atmospheric pressure.

Water Differential Pressure Gages: Water differential pressure gages are used to measure pressure drop across flow meters, terminals, and heat exchangers. A differential pressure gage should be selected so the pressures measured don't exceed the upper limits of the scale. Purge all the air from the gage before reading. Calibration should be every 6 months.

PART 4

Chapter 18
Equations

AIRFLOW

Converting velocity pressure to velocity

$$V = 4005 \sqrt{VP}$$

Where
V = velocity in feet per minute (fpm)
4005 = constant
$\sqrt{VP}$ = square root of the velocity pressure, inches of water gage

Fluid flow equation

$$v = \sqrt{2gh}$$

Where
v = velocity in feet per second
g = gravitational acceleration at 32.2 feet per second squared
h = head, in feet of water

The density of standard air is 0.075 pounds per cubic foot. Substituting in the equation for 1 inch wg of velocity pressure of standard air and making the mathematical computation, results in the constant, 4005.

$$v = \sqrt{2gh}$$
$$v = 60 \sqrt{2 \times 32.2 \times 5.19 \times 13.33}$$
$$v = 4005$$

Where
v = velocity in feet per minute
60 = seconds per minute
5.19 = density of water divided by 12 inches per foot

13.33 = 1 inch of water gage divided by the density of air

Calculating airflow, area and velocity

$Q = AV$

$$A = \frac{Q}{V}$$

$$V = \frac{Q}{A}$$

Where
Q = quantity of airflow in cubic feet per minute (cfm)
A = cross sectional area of the duct in square feet (sf)
V = velocity in feet per minute (fpm)

Calculating air density

$$d = 1.325 \frac{Pb}{T}$$

Where
d = air density in pounds per cubic foot
1.325 = constant, 0.075 divided by 29.92/530
Pb = barometric pressure, inches of mercury
T = absolute temperature (indicated temperature in degrees Fahrenheit plus 460)

Calculating the correction factor for velocity with a change in air density

$$cf = \sqrt{\frac{0.075}{d}}$$

Where
cf = correction factor
0.075 = density of standard air, pounds per cubic foot
d = new calculated density, pounds per cubic foot

Calculating the average velocity corrected for density

$Vc = Vm \times cf$

Where
Vc = corrected velocity
Vm = measured velocity
cf = correction factor for new density

Calculating air volume with a correction for density

$$Q = A \times Vc$$

Where
Q = quantity of airflow in cubic feet per minute
A = area in square feet
Vc = corrected velocity

Calculating air volume, pounds per hour

Lbs per hr = cfm x 4.5

Where
Lbs per hr = pounds per hour of airflow
cfm = quantity of airflow in cubic feet per minute
4.5 = constant, 60 minutes per hour x 0.075 pounds per cubic foot

Air changes per hour and cfm from air changes per hour

$$AC = \frac{cfm \times 60}{Vol}$$

$$cfm = \frac{AC \times Vol}{60}$$

Where
AC = air changes per hour
cfm = quantity of airflow in cubic feet per minute
60 = constant, minutes per hour
Vol = room volume, length x width x height, in cubic feet

AREAS
Rectangular duct

$$A = \frac{ab}{144}$$

Where
A = area of the duct, square feet
a = length of one side of rectangular duct, inches
b = length of adjacent side of rectangular duct, inches
144 = constant, square inches per square foot

Round duct

$$A = \frac{\pi R^2}{144}$$

Where
A = area of the duct, square feet
π = 3.14
R^2 = radius, in inches, squared
144 = constant, square inches per square foot

Flat oval duct

Area of the rectangle plus the area of the circle

Segment of a circle

$$A = \frac{\pi R^2 N}{360}$$

Where
A = area
πR^2 = area of the circle
N = number of degrees in the arc
360 = constant, degrees in a circle

Triangle

$$A = \frac{bh}{2}$$

Where
A = area
b = base of the triangle
h = height of the triangle

BELTS

Belt length

$$L = 2C + 1.57 \, (D + d) + \frac{(D - d)^2}{4C}$$

Where
L = belt pitch length
C = center-to-center distance of the shafts
D = pitch diameter of the large sheave
d = pitch diameter of the small sheave

1.57 = constant, $\dfrac{\pi}{2}$

CIRCULAR EQUIVALENTS OF RECTANGULAR DUCT

Circular equivalent for rectangular duct

$$d = \sqrt{\frac{4 \, ab}{\pi}}$$

Circular equivalent for rectangular duct for equal friction and capacity

$$d = 1.30 \, \frac{ab^{0.625}}{(a + b)^{0.25}}$$

Where
d = equivalent duct diameter
a = length of one side of rectangular duct, inches
b = length of adjacent side of rectangular duct, inches
$\pi = 3.14$

DRIVES

Drive equation

$$RPM_m \times D_m = RPM_f \times D_f$$

$$RPM_m = \frac{RPM_f \times D_f}{D_m}$$

$$D_m = \frac{RPM_f \times D_f}{RPM_m}$$

$$RPM_f = \frac{RPM_m \times D_m}{D_f}$$

$$D_f = \frac{RPM_m \times D_m}{RPM_f}$$

Where
RPM_m = speed of the motor shaft
D_m = pitch diameter of the motor sheave
RPM_f = speed of the fan shaft
D_f = pitch diameter of the fan sheave

ELECTRICAL

Brake horsepower, single-phase circuit

$$BHP = \frac{V \times A \times Eff \times PF}{746}$$

Brake horsepower, three-phase circuit

$$BHP = \frac{V \times A \times Eff \times PF \times 1.73}{746}$$

Where
BHP = brake horsepower
V = volts (For three-phase circuits, this is average volts.)
A = amps (For three-phase circuits, this is average amps.)
Eff = motor efficiency
PF = power factor
746 = constant, watts per horsepower
1.73 = constant, square root of 3, for three-phase circuits

Calculating brake horsepower using no-load amps
(For motors of 10 horsepower and larger.)

$$BHP = \frac{RLA - 0.5\ NLA}{FLA_c - 0.5\ NLA} \times HP_n$$

Calculating field corrected full load amps

$$FLA_c = \frac{V_n \times FLA_n}{V_m}$$

Where
RLA = running load amps, field measured
NLA = no load amps (motor sheave in place, belts removed)
FLA_c = full load amps, field corrected
HP_n = nameplate horsepower
V_n = nameplate volts
FLA_n = nameplate full load amps
V_m = volts, field measured

Single-phase power factor

$$PF = \frac{W}{VA}$$

Three-phase power factor

$$PF = \frac{W}{VA \times 1.73}$$

Where
PF = power factor
W = watts
V = volts
A = amps
1.73 = constant, square root of 3, for three-phase circuits

Voltage unbalance equation

$$\%V = \frac{\Delta D_{max}}{V_{avg}} \times 100$$

$\%V$ = % voltage unbalance (should not exceed 2%)
ΔD_{max} = maximum deviation from average voltage
V_{avg} = average voltage

Current unbalance equation

$$\%C = \frac{\Delta D_{max}}{C_{avg}} \times 100$$

$\%C$ = % current unbalance (should not exceed 10%)
ΔD_{max} = maximum deviation from average amps
C_{avg} = average amps

FANS

Air Horsepower

$$AHP = \frac{CFM \times P}{6356}$$

Fan Brake Horsepower

$$BHP = \frac{CFM \times P}{6356 \times Eff}$$

$$BHP = \frac{CFM \times FSP}{6356 \times FSE}$$

$$BHP = \frac{CFM \times FTP}{6356 \times FTE}$$

Fan Efficiency

$$FE = \frac{CFM \times P}{6356 \times BHP}$$

Fan Static Efficiency

$$FSE = \frac{CFM \times FSP}{6356 \times BHP}$$

Fan Total Efficiency

$$FTE = \frac{CFM \times FTP}{6356 \times BHP}$$

Where
AHP = air horsepower
CFM = airflow volume,
  cubic feet per minute
P = fan pressure, in. wg.
6356 = constant, 33,000 ft-lb/
  min divided by 5.19

Eff = fan efficiency, percent
BHP = brake horsepower
FSP = fan static pressure, in. wg
FSE = fan static efficiency, percent
FTP = fan total pressure, in. wg
FTE = fan total efficiency, percent

Kilowatt usage

$$kW = \frac{CFM \times P}{8520 \times Eff_m \times Eff_f}$$

Where
kW = kilowatts
CFM = airflow volume, cubic feet per minute
P = fan pressure, in. wg
8520 = constant, 6356/.746
Eff_m = motor efficiency, percent
Eff_f = fan efficiency, percent

Temperature rise through the fan (Motors out of the air stream)

$$TR = TSP \times \frac{0.371}{Eff_f}$$

Where
TR = temperature rise through the fan
TSP = static pressure rise through the fan
0.371 = constant, 2545/(6356 × 1.08)
Eff_f = fan efficiency, percent

Temperature rise through the fan (Motors in the air stream)

$$TR = TSP \times \frac{0.371}{Eff_f \times Eff_m}$$

Where
TR = temperature rise through the fan
TSP = static pressure rise through the fan
0.371 = constant, 2545/(6356 × 1.08)
Eff_f = fan efficiency, percent
Eff_m = motor efficiency, percent

Tip Speed

$$TS = \frac{\pi \times D \times RPM}{12}$$

Where
TS = tip speed in feet per minute
D = fan wheel diameter in inches
RPM = revolutions per minute of the fan
π = 3.14
12 = constant, inches per foot

FAN LAWS

Where
CFM_1 = original volume of airflow in cubic feet per minute
CFM_2 = new volume of airflow in cubic feet per minute
RPM_1 = original fan speed in revolutions per minute
RPM_2 = new fan speed in revolutions per minute
Pd_1 = original pitch diameter of the motor sheave
Pd_2 = new pitch diameter of the motor sheave
SP_1 = original static pressure in inches of water column
SP_2 = new static pressure in inches of water column
BHP_1 = original brake horsepower
BHP_2 = new brake horsepower
AMP_1 = original amperage
AMP_2 = new amperage
d_1 = original density in pounds per cubic foot
d_2 = new density in pounds per cubic foot

Fan Law No. 1 Air volume varies in direct proportion to fan speed

$$\frac{CFM_2}{CFM_1} = \frac{RPM_2}{RPM_1}$$

$$RPM_2 = RPM_1 \times \frac{CFM_2}{CFM_1}$$

$$CFM_2 = CFM_1 \times \frac{RPM_2}{RPM_1}$$

Fan Law No. 1 Air volume varies in direct proportion to pitch diameter of the motor sheave

$$\frac{CFM_2}{CFM_1} = \frac{Pd_2}{Pd_1}$$

Fan Law No. 1 Fan speed varies in direct proportion to pitch diameter of the motor sheave

$$\frac{RPM_2}{RPM_1} = \frac{Pd_2}{Pd_1}$$

Fan Law No. 2 Static pressure varies as the square of the fan speed

$$\frac{SP_2}{SP_1} = \left(\frac{RPM_2}{RPM_1}\right)^2$$

$$SP_2 = SP_1 \times \left(\frac{RPM_2}{RPM_1}\right)^2$$

Fan Law No. 2 Static pressure varies as the square of the air volume

$$\frac{SP_2}{SP_1} = \left(\frac{CFM_2}{CFM_1}\right)^2$$

$$SP_2 = SP_1 \times \left(\frac{CFM_2}{CFM_1}\right)^2$$

$$CFM_2 = CFM_1 \times \sqrt{\frac{SP_2}{SP_1}}$$

Fan Law No. 2 Static pressure varies as the square of the pitch diameter

$$\frac{SP_2}{SP_1} = \left(\frac{Pd_2}{Pd_1}\right)^2$$

Fan Law No. 3 Brake horsepower varies as the cube of the fan speed
(For motors of 10 horsepower and larger)

$$\frac{BHP_2}{BHP_1} = \left(\frac{RPM_2}{RPM_1}\right)^3$$

$$BHP_2 = BHP_1 \times \left(\frac{RPM_2}{RPM_1}\right)^3$$

$$RPM_2 = RPM_1 \times \sqrt[3]{\frac{BHP_2}{BHP_1}}$$

Fan Law No. 3 Brake horsepower varies as the cube of the air volume

$$\frac{BHP_2}{BHP_1} = \left(\frac{CFM_2}{CFM_1}\right)^3$$

$$BHP_2 = BHP_1 \times \left(\frac{CFM_2}{CFM_1}\right)^3$$

$$CFM_2 = CFM_1 \times \sqrt[3]{\frac{BHP_2}{BHP_1}}$$

Fan Law No. 3 Brake horsepower varies as the cube of the pitch diameter of the motor sheave

$$BHP_2 = BHP_1 \times \left(\frac{Pd_2}{Pd_1}\right)^3$$

$$Pd_2 = Pd_1 \times \sqrt[3]{\frac{BHP_2}{BHP_1}}$$

Fan Law No. 3 Amperage varies as the cube of the fan speed
(For motors of 10 horsepower and larger)

$$\frac{AMP_2}{AMP_1} = \left(\frac{RPM_2}{RPM_1}\right)^3$$

$$AMP_2 = AMP_1 \times \left(\frac{RPM_2}{RPM_1}\right)^3$$

$$RPM_2 = RPM_1 \times \sqrt[3]{\frac{AMP_2}{AMP_1}}$$

Fan Law No. 3 Amperage varies as the cube of the air volume

$$\frac{AMP_2}{AMP_1} = \left(\frac{CFM_2}{CFM_1}\right)^3$$

$$AMP_2 = AMP_1 \times \left(\frac{CFM_2}{CFM_1}\right)^3$$

$$CFM_2 = CFM_1 \times \sqrt[3]{\frac{AMP_2}{AMP_1}}$$

Fan Law No. 3 Amperage varies as the cube of the pitch diameter of the motor sheave

$$AMP_2 = AMP_1 \times \left(\frac{Pd_2}{Pd_1}\right)^3$$

$$Pd_2 = Pd_1 \times \sqrt[3]{\frac{AMP_2}{AMP_1}}$$

Brake horsepower varies as the square root of the static pressures cubed.

$$BHP_2 = BHP_1 \times \sqrt{\left(\frac{SP_2}{SP_1}\right)^3}$$

FAN LAWS AND DENSITY

Air volume remains constant with changes in air density.

A fan is a constant volume machine and will handle the same airflow regardless of air density. It must be remembered, however, that many instruments are calibrated for standard air density (70 degrees at 29.92 in. Hg) and any change in air density will require a correction factor for the instrument. See "calculating the correction factor for velocity with a change in air density."

Static pressure and brake horsepower vary in direct proportion to density.

$$SP_2 = SP_1 \times \frac{d_2}{d_1}$$

$$BHP_2 = BHP_1 \times \frac{d_2}{d_1}$$

HEAT TRANSFER — AIR

Sensible heat

$$Btuh = CFM \times 1.08 \times TD$$

$$CFM = \frac{Btuh}{1.08 \times TD}$$

$$TD = \frac{Btuh}{1.08 \times CFM}$$

Where

Btuh = Btu per hour sensible heat (heating coil or dry cooling coil load or room load)

CFM = cubic feet per minute volume of airflow

1.08 = constant, 60 min/hr x 0.075 lb/cu ft x 0.24 Btu/lb/F

TD = dry-bulb temperature difference of the air entering and leaving the coil. In applications where cfm to the conditioned space needs to be calculated the TD is the difference between the supply air temperature dry bulb and the room temperature dry bulb.

Latent heat

$$Btuhl = CFM \times 4.5 \times \Delta hl$$

$$\Delta hl = \frac{Btuhl}{4.5 \times CFM}$$

$$CFM = \frac{Btuhl}{4.5 \times \Delta hl}$$

Where
Btuhl = Btu per hour latent heat
CFM = cubic feet per minute volume of airflow
4.5 = constant, 60 min/hour x 0.075 lb/cu ft
Δhl = change in latent heat content of the supply air, btu/lb (from dew point and table of properties of mixtures of air and saturated water vapor)

Total heat

Btuht = CFM x 4.5 x Δht

$$\Delta ht = \frac{Btuht}{4.5 \times CFM}$$

$$CFM = \frac{Btuht}{4.5 \times \Delta ht}$$

Btuht = Btu per hour total heat (a wet cooling coil)
CFM = cubic feet per minute volume of airflow
4.5 = constant, 60 min/hour x 0.075 lb/cu ft
Δht = change in total heat content of the supply air, btu/lb (from wet-bulb temperatures and psychrometric chart or table of properties of mixtures of air and saturated water vapor)

HEAT TRANSFER – WATER

Heat transfer

Btuh = GPM x 500 x ΔT_W

Where
Btuh = Btu per hour, water
GPM = water volume in gallons per minute
500 = constant, 60 min/hour x 8.33 lbs/gallon x 1 Btu/lb/F
ΔT_W = temperature difference between the entering and leaving water

MOTORS
Percent of slip of induction motors and synchronous speed

$$\% = \frac{RPM_S - RPM_r}{RPM_S} \times 100$$

$$RPM_S = \frac{120f}{p}$$

Where
% = percent of slip
RPM_S = synchronous speed in revolutions per minute
RPM_r = rotor speed in revolutions per minute
120 = constant
f = frequency
p = number of poles (not pairs of poles)

PSYCHROMETRICS
Moisture, grains per hour and pounds per hour

Gr per hr = CFM x 4.5 x Δg

$$Lbs\ per\ hr = \frac{Gr\ per\ hour}{7000}$$

Where
Gr per hr = grains per hour, moisture content
CFM = cubic feet per minute volume of airflow
4.5 = constant, 60 min/hour x 0.075 lb/cu ft
Δg = change in moisture content, grains per pound
Lbs per hr = pounds per hour, moisture content
7000 = grains in one pound

Sensible heat ratio

$$SHR = \frac{Hs}{Ht}$$

$$SHR = \frac{0.24\ TD}{\Delta ht}$$
Where
SHR = sensible heat ratio

Hs = Btu per hour sensible heat

Ht = Btu per hour total heat

0.24 = constant, specific heat of air, btu/lb/F

TD = Temperature Difference of the air, dry bulb

Δht = change in total heat content of the supply air, btu/lb

PUMPS

Water horsepower

$$WHP = \frac{GPM \times H \times SpGr}{3960}$$

Pump brake horsepower

$$BHP = \frac{GPM \times H}{3960 \times Eff}$$

$$BHP = \frac{GPM \times TDH}{3960 \times Eff}$$

Pump efficiency

$$Eff = \frac{GPM \times H}{3960 \times BHP}$$

Where

WHP = water horsepower

GPM = water flow rate, gallons per minute

H = pressure (head) against which the pump operates, feet of water

SpGr = specific gravity

3960 = constant, 33,000 ft-lb/min divided by 8.33 lb/gal

BHP = brake horsepower

TDH = total dynamic head, against which the pump operates, feet of
        water

Eff = pump efficiency, percent

Note: For temperatures between freezing and boiling, the specific
gravity is taken as 1.0 and is therefore dropped from the equations
for bhp and efficiency.

Calculating approximate head developed by a centrifugal pump and impeller diameter

$$H = \left(\frac{D \times RPM}{1840}\right)^2$$

$$D = \frac{1840\sqrt{H}}{RPM}$$

Where
H = head in feet of water
D = diameter of the impeller in inches
RPM = revolutions per minute of the impeller
1840 = constant

Kilowatt usage

$$kW = \frac{GPM \times H}{5308 \times Eff_p \times Eff_m}$$

Where
kW = kilowatts
GPM = water volume, gallons per minute
H = pressure, feet of water
5308 = constant, 3960/.746
Eff_m = motor efficiency, percent
Eff_p = pump efficiency, percent

Calculating net positive suction head available

NPSHA = Pa +/− Hs + Hv − Hvp

Where
NPSHA = available net positive suction head expressed in feet
Pa = atmospheric pressure (from altitude table) for elevation of installation, expressed in feet
Hs = gage pressure or vacuum at the suction flange, corrected to pump centerline and expressed in feet (Hs is plus if positive head and minus if vacuum)
Hv = velocity head (from velocity head table) at the point of measurement of Hs, expressed in feet

Hvp = absolute vapor pressure (from vapor pressure/water properties table) of the liquid at pumping temperature, expressed in feet

Example 18.1: A hot water pump is located at sea level. The suction pressure is 12 psig. The velocity is 5 feet per second. The temperature of the water is 160 degrees. Find the NPSHA.

Solving for NPSHA:

NPSHA = Pa + Hs + Hv − Hvp
NPSHA = 33.9′ + 27.72′ + 0.39′ − 10.96′
NPSHA = 51 feet
Pa = atmospheric pressure for sea level, 33.9′
Hs = 12 psig x 2.31 feet/psi = 27.72′

$$H_V = \frac{V^2}{2g}$$

$$H_V = \frac{5^2}{2 \times 32.2}$$

$H_V = 0.39′$
Hvp = from tables = 10.69′

PUMP LAWS

GPM_1 = original volume of water flow in gallons per minute
GPM_2 = new volume of water flow in gallons per minute
RPM_1 = original pump speed in revolutions per minute
RPM_2 = new pump speed in revolutions per minute
D_1 = original diameter of the impeller in inches
D_2 = new diameter of the impeller in inches
H_1 = original head in feet of water
H_2 = new head in feet of water
BHP_1 = original brake horsepower
BHP_2 = new brake horsepower

Pump Law No. 1 Water volume varies in direct proportion to pump speed

$$\frac{GPM_2}{GPM_1} = \frac{RPM_2}{RPM_1}$$

Water volume varies in direct proportion to impeller diameter. Since most HVAC pumps are direct connected, pump impeller diameter will be substituted for rpm in the following pump equations.

$$\frac{GPM_2}{GPM_1} = \frac{D_2}{D_1}$$

$$GPM_2 = GPM_1 \times \frac{D_2}{D_1}$$

$$D_2 = D_1 \times \frac{GPM_2}{GPM_1}$$

Pump Law No. 2 Head varies as the square of the pump speed, gpm or impeller diameter

$$H_2 = H_1 \times \left(\frac{D_2}{D_1}\right)^2$$

$$H_2 = H_1 \times \left(\frac{GPM_2}{GPM_1}\right)^2$$

$$D_2 = D_1 \times \sqrt{\frac{H_2}{H_1}}$$

$$GPM_2 = GPM_1 \times \sqrt{\frac{H_2}{H_1}}$$

Pump Law No. 3 Brake horsepower, or amperage varies as the cube of the pump speed, gpm or impeller diameter
(For motors of 10 horsepower and larger)

$$BHP_2 = BHP_1 \times \left(\frac{D_2}{D_1}\right)^3$$

$$BHP_2 = BHP_1 \times \left(\frac{GPM_2}{GPM_1}\right)^3$$

$$D_2 = D_1 \times \sqrt[3]{\frac{BHP_2}{BHP_1}}$$

$$GPM_2 = GPM_1 \times \sqrt[3]{\frac{BHP_2}{BHP_1}}$$

TEMPERATURE

Temperature conversions
Fahrenheit = 1.8 Celsius + 32
Celsius = (Fahrenheit − 32)/1.8
Rankin = Fahrenheit + 460
Kelvin = Celcius + 273

Log mean temperature difference

$$LMTD = \frac{\Delta T_L - \Delta T_s}{L_n\left(\frac{\Delta T_L}{\Delta T_s}\right)}$$

Where
LMTD = log mean temperature difference
ΔT_L = the larger temperature difference
ΔT_s = the smaller temperature difference
L_n = natural log

Calculating mixed air temperature and percent of outside air

MAT = (%OA X OAT) + (%RA x RAT)

%OA = (MAT − RAT) x 100/(RAT − OAT)

Where
MAT = Mixed Air Temperature
%OA = Percent of Outside Air (cfm)
OAT = Outside Air Temperature

%RA = Percent of Return Air (cfm)
RAT = Return Air Temperature

Calculating fan discharge air temperature and percent of outside air

FDAT = (%OA x OAT) + (%RA x RAT) + 0.5(TSP)

$$\% \, OA = \frac{RAT - [FDAT - 0.5 \, (TSP)]}{RAT - OAT} \times 100$$

Where
FDAT = Fan Discharge Air Temperature
%OA = Percent of Outside Air (cfm)
OAT = Outside Air Temperature
%RA = Percent of Return Air (cfm)
RAT = Return Air Temperature
TSP = Total Static Pressure rise across the fan, in. wg
0.5 = ½ degree per inch of static pressure

Calculating supply air temperature and percent of bypassed air

SAT = (%BA x BAT) + (%CA x CAT)

$$\%BA = \frac{(SAT - CAT) \times 100}{(CAT - BAT)}$$

Where
SAT = Supply Air Temperature
%BA = Percent of Bypassed Air (cfm)
BAT = Bypassed Air Temperature
%CA = Percent of air through the coil (cfm)
CAT = Temperature of the air leaving the coil

Condenser Water Bypass

BCWT = (%BCWR x CWRT) + (%CWS x CWST)

$$\%BCWR = \frac{(BCWT - CWST) \times 100}{(CWRT - CWST)}$$

Where
BCWT = Bypassed Condenser Water Temperature

%BCW = Percent of Bypassed Condenser Water (gpm)
CWST = Condenser Water Supply Temperature
%CWS = Percent of Condenser Water Supply (gpm)
CWRT = Condenser Water Return Temperature

Water Source Heat Pump

CTWT or BRWT = (%RW x RWT) of heat pump X + (%RW x RWT)
of heat pump Y

Where
CTWT = Cooling Tower Water Temperature
BRWT = Boiler Return Water Temperature
RWT = Return Water Temperature
%RW = Percent of Return Water

WATER FLOW
Measuring flow through control valves

$$GPM = C_V\sqrt{\Delta P}$$

$$C_V = \frac{GPM}{\sqrt{\Delta P}}$$

$$\Delta P = \left(\frac{GPM}{C_V}\right)^2$$

Where
GPM = water flow, gallons per minute
ΔP = pressure drop, psi or feet of water
C_V = valve coefficient

Measuring flow through coils
(This equation can only be used when the rated pressure drop is a tested value
by the manufacturer)

$$GPM_2 = GPM_1\sqrt{\frac{\Delta P_2}{\Delta P_1}}$$

$$\Delta P_2 = \Delta P_1 \times \left(\frac{GPM_2}{GPM_1}\right)^2$$

Where
GPM_1 = rated water flow in gallons per minute
GPM_2 = measured or calculated water flow in gallons per minute
ΔP_1 = rated pressure drop, psi or feet of water
ΔP_2 = measured or calculated pressure drop, psi or feet of water

Calculating water flow by heat transfer
(wet cooling coil)

$$GPM = \frac{CFM \times 4.5 \times \Delta ht}{500 \times \Delta T_W}$$

(heating coil)

$$GPM = \frac{CFM \times 1.08 \times \Delta T_A}{500 \times \Delta T_W}$$

Where
GPM = water volume in gallons per minute
500 = constant, 60 min/hour x 8.33 lbs/gallon x 1 Btu/lb/F
Δht = change in total heat content of the supply air, btu/lb (from wet-bulb temperatures and psychrometric chart or table of properties of mixtures of air and saturated water vapor)
ΔT_A = temperature difference between the entering and leaving air
ΔT_W = temperature difference between the entering and leaving water

Velocity head

$$H_V = \frac{V^2}{2g}$$

Where
H_V = velocity head, feet of water
V^2 = velocity of water, feet per second
g = acceleration due to gravity, 32.3 feet per second squared

Velocity correction for pumps when the inlet and outlet connections are different sizes

If the inlet and outlet pipes are the same size, the velocity head is the same entering and leaving and cancel out of the calculation. If however, the pipes are different sizes, there's a velocity head component that's added to the static head rise across the pump to get total dynamic head. However, the velocity head value is usually very small as compared to the static head, and even if it could accurately be plotted on the pump curve, it would make little difference in the test and balance calculations. Therefore, when using the pump as a flow meter the suction and discharge static heads are read and used as the only value to calculate total dynamic head. If, however, corrections are required, some pump manufacturers have a table for velocity corrections. If this table isn't available the equation is:

$$H_V = \frac{V_o^2 - V_i^2}{2g}$$

Where
H_V = velocity head, feet of water
V_o = outlet pipe velocity of water, feet per second
V_i = inlet pipe velocity of water, feet per second
g = acceleration due to gravity, 32.2 feet per second squared

Calculating velocity

$$V = \frac{0.408 GPM}{D^2}$$

V = velocity in feet per second
0.408 = constant, 144 si/sf divided by 60 s/m x 7.5 gal/cf x .25 x 3.14
GPM = water flow, gallons per minute
D = inside diameter of the water pipe, inches

Chapter 19
Conversion Tables

19.1 ALTITUDE PRESSURE TABLE

Altitude in feet	Inches of mercury	Feet of water (SpGr = 1)
−1,000	31.02	
0	29.92	33.9
1,000	28.86	32.8
2,000	27.72	31.6
3,000	26.81	30.5
4,000	25.84	29.4
5,000	24.89	28.2
6,000	23.98	27.3
7.000	23.09	26.2
8,000	22.22	25.2
9,000	21.38	24.3
10,000	20.58	23.4

19.2 GENERAL CONVERSION TABLE

1 Atmosphere equals
33.9 ft. wg
407 in. wg
14.7 psi
29.92 in. Hg

1 boiler horsepower
33,475 Btuh
34.5 pounds of water (evaporated at 970 Btu/lb)

1 Btu equals
778 foot-pounds
0.000393 horsepower-hours
0.000293 kilowatt-hours

1 Btuh
0.000393 horsepower
0.000293 kilowatts

1 cubic foot equals
1728 cubic inches

1 cubic foot of water equals
7.5 gallons

1 cubic foot of water weighs
62.4 pounds

1 ft. wg equals
0.883 in. Hg
12 in. wg
0.433 psi

1 gallon of water equals
231 cubic inches
8.33 pounds

1 horsepower equals
2545 Btuh
42.42 Btu per minute
550 foot-pounds per second
33,000 foot-pounds per minute
0.746 kilowatts
746 watts

1 in. Hg equals
1.13 ft. wg
13.6 in. wg
0.491 psi
70.73 pounds per sq. ft.

1 in. wg equals
0.036 psi
5.2 psf

1 kilowatt equals
3413 Btuh
1.34 horsepower
56.9 Btu per minute

1 mile per hour equals
88 feet per minute

1 pound equals
7000 grains

1 psi equals
2.04 in. Hg
2.31 ft. wg
27.7 in. wg

1 ton of refrigeration equals
12,000 Btuh

1 watt equals
3.41 Btuh
0.00134 horsepower
44.26 foot-pounds per minute

1 year equals
8760 hours
4620 hours of light
4140 hours of darkness

19.3 WATER PROPERTY TABLE

Degrees -F	Density #/CF	Weight #/GAL	Vapor Pressure In Feet Absolute
32	62.40	8.34	
60	62.35	8.33	0.59
100	62.00	8.29	2.20
150	61.20	8.18	8.74
200	60.13	8.04	27.60
212	59.80	7.99	34.00

19.4 TABLE OF CEFAPP RULES OF THUMB, FACTORS, ETC.

1. Recommended air velocities

Offices
   main duct 1200 fpm
   branch duct 800 fpm

Louvers
   intake 400 fpm
   exhaust 500 fpm

Filters
   Fiber
      viscous 700
      dry type 750 fpm
      HEPA 250 fpm
   Renewable
      viscous 500 fpm
      dry 200 fpm
   Electronic
      500 fpm

Coils
      hot water 700 fpm
      cooling 600 fpm

2. Anemometer correction factors
a. If Ak factor for grille is unknown use 0.85
b. For a coil use 0.70

3. Approximate belt length with sheave change

$L_n = 1.57(\Delta d)$

Where

L_n = approximate length of new belt

1.57 = constant, $\dfrac{\pi}{2}$

(Δd) = difference between old sheave pitch diameter and new sheave pitch diameter

4. Belt widths

A 1/2"
B 21/32"
C 7/8"
D 1¼"
E 1½"
2L 9/32"
3L 3/8"
4L ½"
5L 21/32"

5. A CEFAPP rule of thumb for determining barometric pressure is: 0.1 inch Hg reduction from 30" (29.92 rounded off) for each 100' above sea level.

6. A CEFAPP rule of thumb for calculating the velocity correction factor for a change in air density resulting from changes in temperature and altitude is:
a. +2% correction for each 1,000' above sea level.
b. + or − 1% correction for each 10 degrees above or below 70 degrees Fahrenheit.

7. Centrifugal fans
a. If fan efficiency is unknown use 80%.

8. Centrifugal pumps
a. If pump efficiency is unknown use 70%.

9. CFM
a. 1 cfm/sf = 7.5 AC/hr with an 8' ceiling
b. 0.9 cfm/sf = 6 AC/hr with a 9' ceiling
c. 400 cfm per ton of cooling
d. Average cooling load is equal to 1 cfm/sf
e. Equation to find approximate round duct size to carry cfm at 0.1"/100' is duct diameter = $0.90 \ cfm^{0.40}$

10. Heat gain from occupants
a. Moderately active office — 450 Btuh per person (225 sensible and 225 latent)

11. Induction motors
a. Motor efficiency and power factor curves remain fairly flat until the motor load falls below 50%:
    1. Motor efficiency remains between 82% and 92%
    2. Power factor remains between 80% and 90%
b. Synchronous speeds for motors at 60 Hz

Poles	Speed
2	3600
4	1800
6	1200
8	900

c. Approximate amperage ratings

	three-phase		single-phase	
Hp	230V	460V	115V	230V
½	2.0	1.0	9	4.5
¾	2.8	1.4	12	6
1	3.6	1.8	15	7.5
5	15	7.5	25	12.5
10	25	12.5	50	25
25	60	30		
50	120	60		
100	240	120		
200	480	240		

12. Outside air requirements – offices
a.   10% of total air
b.   10-25 cfm per person

13. Correction factor for pressure losses through various types of ductwork (pressure loss times correction factor)

Material	Correction factor
galvanized duct	1.00
fiberglass duct	1.35
lined duct	1.08 - 1.42
flex duct, fully extended	1.85
flex duct, compressed 10%	3.65

14. Pressures
a.   1 in. Hg = 1 ft of water
b.   1 ft. wg = 0.5 psi
c.   1 in. Hg = 0.5 psi

15. Refrigeration
a.

Water TD	gpm/ton
8	3
10	2.4
12	2
20	1.2

b.   hp/ton = 4.71 divided by COP
c.   A COP of 3.5 = 1.34 hp/ton = 1 kW/ton
d.   Condenser tonnage

$$1. \text{ Tons} = \frac{\text{chiller gpm x chiller TD x 1.25}}{24}$$

$$2. \text{ Tons} = \frac{\text{condenser gpm x condenser TD}}{24}$$

e.   Chiller tonnage

$$1. \text{ Tons} = \frac{\text{chiller gpm x chiller TD}}{24}$$

$$\text{2. Tons} = \frac{\text{condenser gpm x condenser TD}}{30}$$

TABLE 19.5 CONVERSION TABLE OF VELOCITY AND VELOCITY PRESSURES

VELOCITY PRESSURE TO VELOCITY

V = velocity in feet pre minute
VP = velocity pressure in inches of water, standard air density

VP	V	VP	V	VP	V	VP	V	VP	V
.01''	400.5	.37	2436	.73	3422	1.09	4181	1.45	4823
.02	566.4	.38	2469	.74	3445	1.10	4200	1.46	4840
.03	693.7	.39	2501	.75	3468	1.11	4219	1.47	4856
.04	801.0	.40	2533	.76	3491	1.12	4238	1.48	4873
.05	895.5	.41	2563	.77	3514	1.13	4257	1.49	4889
.06	981	.42	2595	.78	3537	1.14	4276	1.50	4905
.07	1060	.43	2626	.79	3560	1.15	4295	1.51	4921
.08	1133	.44	2656	.80	3582	1.16	4314	1.52	4938
.09	1201	.45	2687	.81	3604	1.17	4332	1.53	4954
.10	1266	.46	2716	.82	3625	1.18	4350	1.54	4970
.11	1328	.47	2746	.83	3657	1.19	4368	1.55	4986
.12	1387	.48	2775	.84	3669	1.20	4386	1.56	5002
.13	1444	.49	2804	.85	3690	1.21	4405	1.57	5018
.14	1498	.50	2832	.86	3709	1.22	4423	1.58	5034
.15	1551	.51	2860	.87	3729	1.23	4442	1.59	5050
.16	1602	.52	2888	.88	3758	1.24	4460	1.60	5066
.17	1651	.53	2916	.89	3779	1.25	4478	1.61	5082
.18	1699	.54	2943	.90	3800	1.26	4495	1.62	5098
.19	1746	.55	2970	.91	3821	1.27	4513	1.63	5114
.20	1791	.56	2997	.92	3842	1.28	4531	1.64	5129
.21	1835	.57	3024	.93	3863	1.29	4549	1.65	5144
.22	1879	.58	3050	.94	3884	1.30	4566	1.66	5160
.23	1921	.59	3076	.95	3904	1.31	4583	1.67	5175
.24	1962	.60	3102	.96	3924	1.32	4601	1.68	5191
.25	2003	.61	3127	.97	3945	1.33	4619	1.69	5206
.26	2042	.62	3153	.98	3965	1.34	4636	1.70	5222
.27	2081	.63	3179	.99	3985	1.35	4653	1.71	5237
.28	2119	.64	3204	1.00	4005	1.36	4671	1.72	5253
.29	2157	.65	3229	1.01	4025	1.37	4688	1.73	5268
.30	2193	.66	3254	1.02	4045	1.38	4705	1.74	5283
.31	2230	.67	3279	1.03	4064	1.39	4722	1.75	5298
.32	2260	.68	3303	1.04	4084	1.40	4739	1.76	5313
.33	2301	.69	3327	1.05	4103	1.41	4756	1.77	5328
.34	2335	.70	3351	1.06	4123	1.42	4773	1.78	5343
.35	2369	.71	3375	1.07	4142	1.43	4790	1.79	5359
.36	2403	.72	3398	1.08	4162	1.44	4806	1,80	5374

VP	V	VP	V	VP	V	VP	V	VP	V
1.81	5388	2.30	6074	2.79	6690	3.28	7253	3.77	7776
1.82	5403	2.31	6087	2.80	6702	3.29	7264	3.78	7787
1.83	5418	2.32	6100	2.81	6714	3.30	7275	3.79	7797
1.84	5433	2.33	6113	2.82	6725	3.31	7286	3.80	7807
1.85	5447	2.34	6126	2.83	6737	3.32	7297	3.81	7817
1.86	5462	2.35	6139	2.84	6749	3.33	7308	3.82	7827
1.87	5477	2.36	6152	2.85	6761	3.34	7319	3.83	7838
1.88	5491	2.37	6165	2.86	6773	3.35	7330	3.84	7848
1.89	5506	2.38	6179	2.87	6785	3.36	7341	3.85	7858
1.90	5521	2.39	6191	2.88	6797	3.37	7352	3.86	7868
1.91	5535	2.40	6204	2.89	6809	3.38	7363	3.87	7879
1.92	5550	2.41	6217	2.90	6820	3.39	7374	3.88	7889
1.93	5564	2.42	6230	2.91	6832	3.40	7385	3.89	7899
1.94	5579	2.43	6243	2.92	6844	3.41	7396	3.90	7909
1.95	5593	2.44	6256	2.93	6855	3.42	7406	3.91	7919
1.96	5608	2.45	6269	2.94	6867	3.43	7417	3.92	7929
1.97	5623	2.46	6281	2.95	6879	3.44	7428	3.93	7940
1.98	5637	2.47	6294	2.96	6890	3.45	7439	3.94	7950
1.99	5651	2.48	6307	2.97	6902	3.46	7450	3.95	7960
2.00	5664	2.49	6319	2.98	6913	3.47	7460	3.96	7970
2.01	5678	2.50	6332	2.99	6925	3.48	7471	3.97	7980
2.02	5692	2.51	6345	3.00	6937	3.49	7482	3.98	7990
2.03	5706	2.52	6358	3.01	6948	3.50	7493	3.99	8000
2.04	5720	2.53	6370	3.02	6960	3.51	7503	4.00	8010
2.05	5734	2.54	6383	3.03	6971	3.52	7514		
2.06	5748	2.55	6395	3.04	6983	3.53	7525		
2.07	5762	2.56	6408	3.05	6994	3.54	7535		
2.08	5776	2.57	6420	3.06	7006	3.55	7546		
2.09	5790	2.58	6433	3.07	7017	3.56	7556		
2.10	5804	2.59	6445	3.08	7028	3.57	7567		
2.11	5817	2.60	6458	3.09	7040	3.58	7578		
2.12	5831	2.61	6470	3.10	7051	3.59	7588		
2.13	5845	2.62	6482	3.11	7063	3.60	7599		
2.14	5859	2.63	6495	3.12	7074	3.61	7610		
2.15	5872	2.64	6507	3.13	7085	3.62	7610		
2.16	5886	2.65	6519	3.14	7097	3.63	7630		
2.17	5899	2.66	6532	3.15	7108	3.64	7641		
2.18	5913	2.67	6544	3.16	7119	3.65	7652		
2.19	5927	2.68	6556	3.17	7131	3.66	7662		
2.20	5940	2.69	6569	3.18	7142	3.67	7672		
2.21	5954	2.70	6581	3.19	7153	3.68	7683		
2.22	5967	2.71	6593	3.20	7164	3.69	7693		
2.23	5981	2.72	6605	3.21	7176	3.70	7704		
2.24	5994	2.73	6617	3.22	7186	3.71	7714		
2.25	6008	2.74	6629	3.23	7198	3.72	7724		
2.26	6021	2.75	6641	3.24	7209	3.73	7735		
2.27	6034	2.76	6654	3.25	7220	3.74	7745		
2.28	6047	2.77	6666	3.26	7231	3.75	7755		
2.29	6060	2.78	6678	3.27	7242	3.76	7766		

19.6 TABLE OF ASSOCIATIONS AND ORGANIZATIONS

AABC: Associated Air Balance Council.

ACGIH: American Conference of Governmental Industrial Hygienists.

ADF: Air diffusion Council.

AMCA: Air Movement and Control Association.

ASHRAE: American Society of Heating, Refrigerating and Air-Conditioning Engineers.

CTI: Cooling Tower Institute

NBS: National Bureau of Standards.

NEBB: National Environmental Balancing Bureau.

NFPA: National Fire Protection Association.

SMACNA: Sheet Metal & Air Conditioning Contractors' National Association.

INDEX